The F100 Engine Purchasing and Supply Chain Management Demonstration

Findings from Air Force Spend Analyses

Mary E. Chenoweth, Clifford A. Grammich

T0150151

Prepared for the United States Air Force

PROJECT AIR FORCE

The research reported here was sponsored by the United States Air Force under Contract F49642-01-C-0003. Further information may be obtained from the Strategic Planning Division, Directorate of Plans, Hq USAF.

Library of Congress Cataloging-in-Publication Data

Chenoweth, Mary E.
  The F100 engine purchasing and supply chain management demonstration : findings from Air Force spend analyses / Mary E. Chenoweth, Clifford A. Grammich.
    p. cm.
  "MG-424."
  Includes bibliographical references.
  ISBN 0-8330-3889-3 (pbk. : alk. paper)
  1. United States. Air Force—Procurement—Evaluation. 2. Jet engines—United States—Costs. 3. Spare parts—United States—Costs. 4. United States. Air Force—Supplies and stores—Evaluation. I. Grammich, Clifford A. (Clifford Anthony), 1963– II. Title.

UG1123.C45 2006
358.4'183—dc22

                                                                      2006013368

The RAND Corporation is a nonprofit research organization providing objective analysis and effective solutions that address the challenges facing the public and private sectors around the world. RAND's publications do not necessarily reflect the opinions of its research clients and sponsors.

RAND® is a registered trademark.

Published 2006 by the RAND Corporation
1776 Main Street, P.O. Box 2138, Santa Monica, CA 90407-2138
1200 South Hayes Street, Arlington, VA 22202-5050
4570 Fifth Avenue, Suite 600, Pittsburgh, PA 15213-2665
RAND URL: http://www.rand.org/
To order RAND documents or to obtain additional information, contact
Distribution Services: Telephone: (310) 451-7002;
Fax: (310) 451-6915; Email: order@rand.org

Preface

This monograph describes spend analyses that the RAND Corporation conducted in 2002 for Phase I of the Purchasing and Supply Management (PSM) demonstration at the Oklahoma City Air Logistics Center (OC-ALC) for purchases of F100 jet engines and jet engine bearings. As part of the Spares Campaign begun in early 2001 under Air Staff leadership to reengineer Air Force supply, the objective of the PSM demonstration was to apply best practices to managing supplies, suppliers, and the supply base to attain the best quality, performance, and prices in purchased goods and services. RAND provided analytical support to OC-ALC as it established business rules for collecting and analyzing spend data. The RAND analyses were conducted using data from several Air Force and Defense Logistics Agency (DLA) databases.

RAND concluded its participation in the demonstration in October 2002 and transferred its analyses to the Air Force. Soon afterward, the Air Force implemented PSM best practices for all its Air Logistics Center purchases and implemented the Purchasing and Supply Chain Management (PSCM) initiative. PSCM is one of the major transformation initiatives of the Air Force Materiel Command to implement Expeditionary Logistics for the 21st Century. Since the decision to implement PSCM, the Air Force has constructed the Strategic Sourcing Analysis Tool to facilitate routine spend analyses for spares and repair purchases.

This monograph should be of interest to those involved in PSCM-related spend analyses, especially analyses for Air Logistics

Centers, and others with an interest in such analyses. This report is not intended to provide a broad overview of how spend analyses are conducted. Readers interested in further information on this subject should consult earlier RAND research (Moore et al., 2002, and Moore et al., 2004) and other literature cited in this report.

This work was conducted by the Resource Management Program of RAND Project AIR FORCE and was sponsored by the U.S. Air Force Deputy Chief of Staff for Logistics, Installations, and Mission Support, Directorate of Transformation (USAF/A4I) and the Deputy Assistant Secretary (Contracting) (SAF/AQC). It is part of a broader study titled "Air Force Purchasing and Supply Chain Management: Support and Evaluation of the ALC-Wide Demonstrations and the Proposed Organization."

Similar RAND Corporation work for the U.S. Air Force has been documented in the following reports:

- *An Assessment of Air Force Data on Contract Expenditures,* by Lloyd Dixon, Chad Shirley, Laura H. Baldwin, John A. Ausink, and Nancy F. Campbell, MR-274-AF, 2005.
- *Using a Spend Analysis to Help Identify Prospective Air Force Purchasing and Supply Management Initiatives: Summary of Selected Findings,* by Nancy Y. Moore, Cynthia R. Cook, Clifford A. Grammich, and Charles Lindenblatt, DB-434-AF, 2004.
- *Implementing Performance-Based Services Acquisition (PBSA): Perspectives from an Air Logistics Center and a Product Center,* by John A. Ausink, Laura H. Baldwin, Sarah Hunter, and Chad Shirley, DB-388-AF, 2002.
- *Implementing Best Purchasing and Supply Management Practices: Lessons from Innovative Commercial Firms,* by Nancy Y. Moore, Laura H. Baldwin, Frank Camm, and Cynthia R. Cook, DB-334-AF, 2002, www.rand.org/publications/DB/DB334.
- *Federal Contract Bundling: A Framework for Making and Justifying Decisions for Purchased Services,* by Laura H. Baldwin, Frank Camm, and Nancy Y. Moore, RAND MR-1224-AF, 2001.

- *Performance-Based Contracting in the Air Force: A Report on Experiences in the Field*, by John Ausink, Frank Camm, and Charles Cannon, DB-342-AF, 2001.
- *Strategic Sourcing: Measuring and Managing Performance*, by Laura H. Baldwin, Frank Camm, and Nancy Y. Moore, DB-287-AF, 2000.

RAND Project AIR FORCE

RAND Project AIR FORCE (PAF), a division of the RAND Corporation, is the U.S. Air Force's federally funded research and development center for studies and analyses. PAF provides the Air Force with independent analyses of policy alternatives affecting the development, employment, combat readiness, and support of current and future aerospace forces. Research is conducted in four programs: Aerospace Force Development; Manpower, Personnel, and Training; Resource Management; and Strategy and Doctrine.

Additional information about PAF is available on our Web site at http://www.rand.org/paf.

Contents

Figures

Tables

Summary

Purchasing and supply chain management (PSCM) offers the Air Force a means to make better use of its resources in general and to improve several of its logistics processes specifically. Conducting a spend analysis is one of the first steps in implementing PSCM practices. A spend analysis that documents what is purchased, how much is spent, and where goods and services are purchased can help an enterprise to identify specific performance, quality, and cost goals in relationships with providers and can suggest where time and resources should be focused to achieve those goals.

In fiscal year (FY) 2002, the Air Force chose engine parts as an area for examining the feasibility of employing best practices for purchasing and supply management initiatives. Oklahoma City Air Logistics Center (OC-ALC), which is responsible for supporting Air Force engines, then selected the F100 engine as its platform for a PSM demonstration. RAND was asked to assist OC-ALC in conducting a spend analysis on F100 engines, which led to a spend analysis of jet engine bearings, a critical component for jet engine maintenance.

The F100 engine has remained in inventory longer than originally planned and powers more Air Force jet aircraft than any other engine. Because maintaining the F100 and other jet engines constitutes such a large part of Air Force operations, any improvements in purchasing and supply management of jet engines would lead to noticeable improvements in equipment cost and performance throughout the service.

A spend analysis involves an iterative, four-step process—extracting data from the best sources, integrating and validating the data to ensure their accuracy and completeness, cleansing data to eliminate discrepancies in the data, and analyzing the data—with the process repeated as data are improved or as new issues are identified for analysis (see pp. 9–10).

At the time of the Air Force's PSM demonstration, there was no single source of data for the OC-ALC's spend analysis.[1] Instead, data were integrated from a variety of sources. For the spend analysis conducted by RAND, Air Force data from the following sources were used:

- Contract Action Reporting System (J001)
- Acquisition and Due-in System (J041)
- Contract Depot Maintenance and Cost System (G072D)
- Automated Budget Compilation System (ABCS) (D075)
- Item Manager Wholesale Requisition Process (D035A)
- Bill of Materials (BOM) (D200F)
- Contracting Business Intelligence System (CBIS)
- Acquisition Method Code Screening System (J090A).

Because the Defense Logistics Agency (DLA) also purchases goods and services for the F100 engine, we also used data from the following DLA data sets:

- Active Contract File
- Requisition File.

RAND examined Air Force and DLA data for FY 1999–2002. Air Force spending on F100 items during the years studied varied between $439 million and $670 million per year (see p. 27). Our analysis of data indicates that items, i.e., spare parts and repair serv-

---

[1] The Strategic Sourcing Analysis Tool, which the Air Force developed to implement PSCM, brings together information required for spend analyses from many legacy data systems. This study predates the development of this tool.

ices that can be linked to a National Stock Number (NSN), constitute most of Air Force F100 spending (see p. 28). The bulk of other Air Force purchases for this engine, primarily for acquisition and testing of new F100 equipment, could not be linked to an NSN.

Of the F100 items that the Air Force purchased, most were for sustainment of engines (see p. 31). Purchases by Air Logistics Centers (ALCs), which purchase nearly all sustainment items associated with an NSN, were primarily for spare parts. Most ALC F100 contract repair dollars were for a Pratt & Whitney Total Systems Support (TSS) contract for the F100-PW-229 engine, and much of the remaining F100 repair dollars were for contracts to help bridge a workload transition from the San Antonio ALC to the Oklahoma City ALC. This meant that only a small portion of ALC F100 repair purchases could be considered a prospective target for PSCM improvements. Many of these repair purchases were through sole-source contracts, and even "competitive" contracts were almost uniformly limited to qualified sources.

Air Force F100 expenditures were significantly greater than DLA F100 expenditures, which averaged about $102 million a year (see pp. 28 and 40). (This dollar figure likely is an overestimate given the difficulties of isolating DLA F100 spending.) However, Air Force F100 expenditures were concentrated in fewer contracts and NSNs. Although their patterns of concentration of spending among certain numbers of contracts and NSNs differed, both the Air Force and DLA had large portions of their F100 item spend concentrated among a small number of supplier firms (see pp. 40–42). The concentration of spending among top producers suggests that some opportunities to improve PSCM processes with these suppliers exist, including consolidating the number of contracts with those suppliers or exploring other ways to take advantage of their level of spending to gain leverage with top suppliers.

To drive down management costs, both the Air Force and DLA may wish to reduce their total number of suppliers where there are redundant sources of supply, especially for those suppliers with whom they spend relatively few dollars. Such contraction of the supply base, and in the number of required contracts, would (1) free up contracting personnel to become more familiar with the industries with which they work, including best practices in those industries; (2) enable logistics organizations to devote more time to developing strategic relationships with their key suppliers and working on continuous supply-chain improvements; and (3) reduce transaction costs. Both the Air Force and DLA may also wish to consider potential improvements through collaboration, with the agency that purchases more items from a common supplier—typically the Air Force for F100 items—leading the effort to improve PSCM practices.

The results of the RAND analysis demonstrate how a spend analysis for a weapon system can lead to targeting specific items for additional analyses and PSCM initiatives. As stated above, the choices for items that would be the basis of a contract featuring PSCM improvements were limited. A contract with Pratt & Whitney was close to completion at the start of the demonstration and a collaborative effort was under way with DLA to form a strategic supplier alliance with Honeywell International. Jet engine bearings were chosen from among the items that might be appropriate for PSCM initiatives (see pp. 43–44). The Air Force spends millions of dollars on bearings annually and past supply-chain problems with this group of items have adversely affected readiness.

While Air Force purchases of F100 items exceeded DLA's F100 purchases, DLA's purchases of jet engine bearings, which averaged $18.5 million annually, were more than twice the amount of the Air Force's purchases of bearings, which averaged $8.7 million a year (see p. 46). DLA spending was concentrated in spare consumable bearings, whereas Air Force spending was concentrated in more expensive fracture- and safety-critical bearings. Air Force spending for jet engine bearings was also more concentrated in sole-source items (see pp. 48–49).

The Air Force and DLA shared many common suppliers for F100 items, and they shared many common suppliers for jet engine bearings (see p. 54). Among most of these suppliers, DLA spent more for bearings than did the Air Force, but among some of the suppliers, the Air Force had higher total expenditures for all goods and services. While DLA spent more with several bearings suppliers than did the

Air Force, the Air Force or another service had a higher overall average annual spend for goods and services with every bearings supplier. Efforts to increase leverage with suppliers may best be led by the service that spends the most money with those suppliers. Such strategic efforts would not preclude an individual service from having contracts with suppliers that address its specific needs.

Air Force data from the sources listed earlier in this summary can help to identify opportunities for PSCM improvements for both large and relatively small but critically important items. As the Air Force gains more experience in conducting spend analyses, it will undoubtedly uncover further means for getting the most from its resources.

Acknowledgments

The spend analyses described in this document benefited tremendously from the generous backing of the Oklahoma City Air Logistics Center F100 Purchasing and Supply Chain Management demonstration team, whose members provided data and expert knowledge.

We are grateful for the help from many members of the demonstration teams at at OC-ALC, including Col Reginald Banks (Ret), formerly the leader of the F100 Engine PSCM Team at OC-ALC; Darla Bullard, 448 MSUG/GBMOP, who led the Data Team and provided numerous data extracts for RAND; Dewayne Jones, 327 TSG/GFTR, who led the Strategy Development Team for the OC-ALC F100 and jet engine bearings spend analyses; Maj Scott Jones, OC-ALC/CCA, who provided support to Colonel Banks; Bill LaPach, 448 MSUG/GBMOS, who led the Information Technology Team, which provided technical support to OC-ALC demonstration team; Felix Lopez, 76 AMXG/MXAAWS, who led the PSCM Spend Analysis Sub-Team; Janice Moody, 448 MSUG/GBMOP, who led the Supply Chain Mapping Team; and Michael Yort, 448 ACSG/GBCW, who led the Vendor Analysis Sub-Team, which was later incorporated into the Strategy Development Team. We are also very grateful to Glenn Starks, Defense Supply Center—Richmond, who helped secure relevant Defense Logistics Agency data and assisted in their interpretation.

All these individuals provided us with enormous amounts of data on numerous occasions, answered our questions, assisted in several components of the analysis requiring ALC information, and helped us to develop rules for addressing data anomalies. We also thank many others not named above whose efforts helped us indirectly.

Finally, we thank our colleague John Ausink, of RAND, and Mohan Sodhi, of the Cass Business School, London, UK, who provided helpful comments and suggestions on this document.

Acronyms

ABCS	Automated Budget Compilation System
ACF	Active Contract File
ACSG	Aircraft Commodities Sustainment Group
AFKS	Air Force Knowledge System
AFMC	Air Force Materiel Command
AFMC/FM	Air Force Materiel Command Financial Management
ALC	Air Logistics Center
AMC	Acquisition Method Code
AMXG	Aircraft Maintenance Group
ASC	Aeronautical Systems Center
BOM	Bill of Materials
BRAC	Base Realignment and Closure
CBIS	Contracting Business Intelligence System
CCC	Canadian Commercial Corp.
DCMA	Defense Contract Management Agency
DD350	Form DoD 350, Individual Contract Action Report
DLA	Defense Logistics Agency
DLR	depot-level reparable

DoD	Department of Defense
DUNS	Data Universal Numbering System
ELOG21	Expeditionary Logistics for the 21st Century
FSC	Federal Supply Class
FY	fiscal year
GBCW	Accessories Sustainment Squadron
GSA	General Services Administration
MOCAS	Mechanization of Contract Administration Services
NSN	National Stock Number
OC-ALC	Oklahoma City Air Logistics Center
OEM	original equipment manufacturer
PAF	Project AIR FORCE
PSC	Product and Service Code
PSCM	Purchasing and Supply Chain Management
PSM	Purchasing and Supply Management
PW	Pratt & Whitney
RDTE	research, development, and technical evaluation
SAF/AQC	U.S. Air Force Deputy Assistant Secretary (Contracting)
TSG	Tanker Sustainment Group
TSS	Total System Support
UTC	United Technologies Corporation
WSDC	Weapon Systems Designator Code

CHAPTER ONE

Introduction

Like many enterprises in the private sector, the Air Force seeks to make better use of its resources and to improve its logistics and equipment sustainment processes. It has defined its efforts for doing so in its Expeditionary Logistics for the 21st Century (eLog21) plan. ELog21 seeks to increase equipment availability while reducing annual operations and equipment sustainment costs (U.S. Air Force Deputy Chief of Staff Installations and Logistics, 2004). The Air Force Materiel Command (AFMC) has developed several initiatives to implement eLog21, including those for Purchasing and Supply Chain Management (PSCM).

PSCM has its roots in the Spares Campaign initiative of 2001, which sought to improve spares availability and warfighter readiness. The Spares Campaign included Purchasing and Supply Management (PSM) as one of eight initiatives designed to improve weapon system availability by improving spares availability (Mansfield, 2002; Rukin, 2001). The emphasis of the Spares Campaign on PSM coincided with reports of significant performance, quality, and cost improvements that commercial companies were realizing by integrating purchasing in their supply management operations.[1] RAND Corpora-

---

[1] Because the commercial sector refers to such practices as "purchasing and supply management," while the Air Force now refers to those practices as "purchasing and supply chain management," we use both terms somewhat interchangeably in this document. We generally reserve the use of the term PSCM for describing specific Air Force practices (e.g., developing better purchasing practices for engine bearings) designed to integrate the tenets of PSM with the Air Force's supply chain management.

tion research efforts (documented in Moore et al., 2002, and Moore et al., 2004) outlined general principles and practices of the private sector that the Air Force could adapt to its purchasing activities.

To demonstrate the benefits of improved purchasing and supply management in an Air Force setting, the Air Force launched a demonstration of PSM best practices at the Oklahoma City Air Logistics Center (OC-ALC), selecting the F100 engine for the demonstration. The objective of the demonstration was to develop a supply strategy and contract for a group of F100 requirements that would incorporate the tenets and embody the principles of PSM best practices (U.S. Air Force and Oklahoma City Air Logistics Center, 2002 and 2003) as recognized by the research literature and as practiced by innovative enterprises, to attain the best quality, performance, and prices for purchased goods and services. The demonstration had two phases, one examining existing processes that would be affected by implementing PSM best practices and the other examining new processes and improvements that would be necessary for the implementation of such practices.

One of the first steps that leading private enterprises undertake to implement PSM best practices is a spend analysis (Aberdeen Group, 2002). In developing proactive supply strategies for the acquisition and management of a group of purchased goods or services, an enterprise should, ideally, focus first on the goods and services that would have the greatest impact on the performance of the enterprise and for which implementing supply strategies would require less effort and entail lower risk than other sorts of goods and services. A spend analysis helps enterprises to identify prospective targets for improvements by answering such questions as what is purchased, how much is spent, and where goods and services are bought.

This report documents the results of a spend analysis RAND conducted for Phase I of the F100 demonstration of PSM best practices. We begin this chapter by discussing the questions that a spend analysis can help an enterprise to answer and why the Air Force chose the F100 engine for its demonstration of PSM best practices.

What Is a Spend Analysis, and Why Do Enterprises Use It?

Regardless of the target chosen for purchasing and supply management innovations, a spend analysis is a necessary first step for developing a supply strategy. It can help an enterprise to identify and achieve specific performance, quality, and cost goals in its relationships with outside providers. Specifically, enterprises use spend analyses to answer such questions as the following:

- What are we buying? Spend analyses begin with historical information on what an enterprise buys and at what cost. This information may be gathered and analyzed by systems, by commodity groups, or by suppliers.[2]
- Who is buying? The Air Force enterprise includes all goods and services managed by the Air Force, and all purchased Air Force goods and services managed by other military agencies (e.g., the Defense Logistics Agency [DLA]). Information on how the enterprise's expenditures are distributed for specific commodities or suppliers can yield insights for future supply strategies, including efforts to leverage purchases with particular suppliers. The number of contracts and other information on frequency of purchasing for Air Force goods and services also may yield insights for possible contract consolidations that would reduce long-term administrative costs and improve indirect cost efficiencies, particularly with sole-source or best-value suppliers that have many similar contracts with the Air Force.[3]
- Who are our suppliers? Knowing key suppliers can also help an enterprise to identify opportunities for making PSM improve-

---

[2] Spend analyses are concerned solely with direct purchases, not organizational internal costs, such as transaction costs. Nevertheless, spend analyses can identify areas where purchasing practices may be leading to higher transaction costs—e.g., purchases of the same goods or services from multiple suppliers or on multiple contracts.

[3] Consolidating requirements into fewer contract solicitations may actually increase administrative lead times in the short run, given the additional coordination such initial efforts require.

ments. Spend analyses can be used to rank suppliers by total number of contracts and total spending, which can help to identify firms with which an enterprise may seek to develop more strategic relationships. Identifying key suppliers over time can also provide insight into the composition of the supply base (e.g., original equipment manufacturers or third-party suppliers and distributors), and information on mergers and acquisitions could indicate new opportunities to improve purchasing practices with suppliers who have increased their business or have new corporate leadership. Trends in the supply base, such as the selling of business units by suppliers, suppliers changing their products or product mix, increased prices, and vanishing vendors, all indicate potential problems to address in purchasing and supply management.

Spend analyses are often linked to future requirements to inform decisions on how to develop supply strategies for upcoming purchases of goods and services. Specifically, those analyses are often combined with analyses that ask the following:

- What do we need to address in the future? The Air Force can improve supplier strategies if it considers future requirements within the context of its supply base. Estimating future demands and needs across the enterprise allows the Air Force not only to maximize its leverage and develop strategic relationships with key suppliers but also to better manage its supply base and potential risks (resulting, for example, from limited competition or low or variable demand for a commodity).
- Where and how much could we improve? The information in spend analyses on the relative importance of varying suppliers and commodities, when combined with information on past performance, provides insight on specific areas the Air Force may wish to target in purchasing and supply management initiatives, including improvements in availability, quality, and cost containment.

- How should we manage the supply base? Spend analyses can yield information on how to develop flexible supply bases that will help the Air Force to better cope with inevitable uncertainties. Such analyses also help guide efforts for continuous improvements in relationships with key suppliers, particularly through improved knowledge of industry-wide advances. Increasing numbers of aircraft operating beyond their originally planned life spans makes active management of the supply base particularly important. Equipment aging can lead to new support issues, including unpredictable performance of advanced-age materials or systems. For example, there may not be parts or suppliers available to replace parts or systems that fail for the first time after extended service. In some cases, a spend analysis may help indicate that developing a repair system for certain consumable parts would be more economical than maintaining a supply of such parts.

Enterprises conduct spend analyses for three reasons. First, they can demonstrate to senior leadership how purchasing and supply management initiatives can help to achieve other goals, particularly goals related to reducing overall costs or having financial resources available for other elements of the enterprise. This is commonly known as "making the business case for change." Spend analyses may, for example, illustrate purchasing inefficiencies resulting from a large number of contracts with a single supplier or product, or a large number of buying offices purchasing similar goods and services in isolation rather than working together to increase their purchasing leverage. Combining such purchasing efforts, and increasing contract efficiencies, can improve the overall efficiency of enterprise expenditures.

Second, a spend analysis can help managers target specific commodity groups and specific items within those groups for PSM initiatives. The spend analysis can also help determine which organizations ought to lead purchasing and supply management initiatives and which others ought to be included in the effort. Typically, spend analyses have led commercial firms to first target those commodities that will yield high and rapid rewards and that have little risk or difficulty in implementation of PSM initiatives and to then focus on others that pose more risk or difficulty for yielding savings.

Third, a spend analysis, when conducted on an ongoing basis, can help managers develop new supply strategies. Updated knowledge of areas where purchasing leverage may be improved can be a continual help in achieving other enterprise goals.

Applying a Spend Analysis to F100 Engine Support

As mentioned above, the Air Force chose F100 engine support as a target for the PSM demonstration. The F100 is of considerable importance to Air Force operations. The engine, manufactured by Pratt & Whitney, is the only engine for F-15 fighters and the engine for more than two-thirds of F-16 aircraft worldwide (Grimes, 2003; Pratt & Whitney, 2005). The F100 powers more Air Force jets than any other engine. The Air Force has nearly 3,300 F100 engines, worth approximately $11.6 billion (U.S. Air Force and Oklahoma City Air Logistics Center Supply Chain Transformation Team, 2003).

The F100 engine has remained in the Air Force's inventory longer than originally planned and powers more Air Force jet aircraft than any other engine. In 2001, about 6 percent of the F100 engines and major modules were nearing the end of their originally designed service life and were on their third and last interval, having been overhauled twice at an Air Force depot (Dues, 2001; Grimes, 2003). By 2010, the proportion of engines and major modules operating in their third interval is expected to increase to 97 percent. Consequently, without upgrades, OC-ALC anticipates that these engines increasingly will have higher maintenance costs and parts shortages (U.S. Air Force and Oklahoma City Air Logistics Center Supply Chain Transformation Team, 2003). This situation has created some of the most significant cost and readiness issues that the Air Force faces. In fiscal year (FY) 2002, the Air Force spent $899 million on acquiring or maintaining F100 engines; 24 percent of all expenditures for acquiring and supporting engines and 44 percent of all expenditures on engine sustainment were for the F100.[4]

The F100 is such an important part of Air Force engine expenditures, and jet engines and their maintenance constitute such a large part of Air Force operations, that any improvements achieved in purchasing and supply management for the F100 would lead to noticeable improvements in cost and performance throughout the service.

The results presented here demonstrate how a spend analysis for a weapon system can lead to targeting of more specific items for additional analyses and initiatives. The research regarding the F100 engine offers a weapon system perspective on the use of a spend analysis, including management of multiple suppliers who each contribute system-specific goods and services. Among the commodities needed for F100 engine maintenance, the Air Force selected for further analysis acquisition of jet engine bearings, a product whose availability can ultimately affect the timeliness of engine maintenance and repair.

Spend analyses of a weapon system and of a specific commodity for that system also help to highlight the dual perspectives of such analyses. From one perspective, the Air Force needs to align its purchase of goods and services for weapon systems, engines, and engine modules to meet the demands of specific customers (e.g., Major Commands). This approach helps to ensure that the service meets its requirements for combat capability. From another perspective, the Air Force, as discussed above, needs to determine the specific goods and services it is purchasing to meet those requirements and needs to identify all the relationships it is having with its providers.[5] This approach could be particularly helpful in reducing the number of con-

---

[4] The Air Force Contract Action Reporting System (J001), which records DD350 data, indicates that in FY 2002 the Air Force spent $2.1 billion on gas turbines and jet engine aerospace components. That amount was greater than that spent on any other group of components or systems except for the $10.6 billion spent on fixed-wing aircraft. The monies spent on the F100 far exceeded the $291 million spent on the F110 engine that also powers F-16 aircraft.

[5] Such a perspective is typical of commodity councils in private industry that develop and execute tailored supply strategies for specific goods and services (Savoie, 2003).

tracts and identifying which supplier relationships to manage strategically. This perspective is also likely to include purchases of a given commodity for multiple weapon systems.

The required level of detail and level of accuracy of the data used for a spend analysis are determined by the purpose of the analysis. For discussion of service-wide or strategic purchasing issues, such as making the case to change purchasing practices, highly aggregated information can suffice. For tactical matters, such as implementing new PSCM practices into specific contracts, more refined data are necessary. While highly accurate data for spend analyses may prove to be economically impractical to compile, the data should be sufficiently accurate so that improvements in their accuracy will not affect the conclusions that are drawn from them.[6] Ensuring such accuracy will require greater effort for more tactical analyses.

Organization of This Report

In the next chapter, we examine in more detail the data and process needed for a spend analysis. In Chapter Three, we review our results of a spend analysis for the F100 engine for FY 1999 through FY 2002 and show categories of items purchased and top suppliers in terms of spend.[7] In Chapter Four, we review our results of a spend analysis for jet engine bearings for FY 1999–2002 and also show top suppliers in terms of spend. In Chapter Five, we review the implications of our research for F100 PSCM and for future spend analyses.

---

[6] In interviews with RAND, representatives from private-sector firms said that the added expense of ensuring data accuracy is more than made up by the savings from making better decisions.

[7] In this report, the word "spend" is often used to refer to the purchase of goods and services. This usage of the term reflects its usage in the PSM best practices literature.

CHAPTER TWO

Spend Analysis Methods and Data

Enterprise-wide spend analyses require the extraction, integration, and analysis of all spend data from business units. These spend data often are obtained from legacy systems developed for various purposes and do not necessarily contain all the relevant information needed for a spend analysis. Spend analyses require data on all purchased goods and services, contracts, and suppliers. Private-sector companies have reported that their initial spend analyses often do not include all spend data, but over time, their analyses became more complete as additional spend data are identified, extracted, integrated, and analyzed (see Verespej, 2005; Porter et al., 2004).

Spend analyses involve a multistep process. The first step, extraction, requires all relevant spending data across an enterprise's multiple business units. These business units, like the Air Force, may have various legacy systems, which typically are designed for purposes other than collecting data for a spend analysis and that record similar data elements in differing formats. Spend data should, if possible, be extracted from original sources to ensure the best-quality data possible. When data are extracted from multiple systems, they need to be synthesized. This leads to the second step, integration and validation of data by experts. Ensuring accuracy and completeness of data leads to the third step, data cleansing to eliminate discrepancies. If data are inaccurate and/or incomplete, and if such problems could affect decisionmaking outcomes, steps must be taken to remedy those problems (for more information on the implications of data quality problems and what to do about them, see Kanakamedala, Ramsdell, and

Roche, 2003). After data have been extracted, integrated, validated, and cleansed, they then can be analyzed. These steps can be repeated as more is learned about the data and as new data become available. In particular, analyses can uncover problems with the data that may require repeating the four-step process as more cleansed or refined data become available or as additional required data are identified (Minahan and Vigoroso, 2003). Some leading commercial firms have dedicated staff whose responsibility is to gather and cleanse data for spend analyses and inventory management.

Extracting and Integrating Relevant Data

For the F100 spend analysis, we integrated data provided to us by the OC-ALC F100 Implementation Team from nearly a dozen sources, including the Air Force, DLA, and the Defense Contract Management Agency (DCMA). Many of these data can now be accessed through the Air Force Knowledge System (AFKS), a data warehouse of some Air Force legacy systems, and specifically through the Strategic Sourcing Analysis Tool within AFKS.[1] In this section, we review each of the data sources that we used.[2]

Data from the Air Force included those from the following sources:

---

[1] The AFMC Strategic Sourcing Analysis Tool integrates the data contained in the legacy systems described in this report with data from other systems. The tool was developed to support spend analyses required by the AFMC Commodity Councils to develop supply strategies for their specific commodity group requirements. The analysis tool contains detailed item data related to spend, forecasts, requisitions, MICAPs, back-orders, and lead times. It was developed after the F100 demonstration and was not available for the F100 PSM demonstration spend analyses.

[2] One of our reasons for providing details on these data is to demonstrate the novelty of a spend analysis for Air Force purposes and the issues that the Air Force confronted in conducting initial spend analyses. Other services are dealing with similar issues as they conduct spend analyses both service-wide and for particular weapon systems and commodities. The difficulties encountered in such analyses highlight the need for an analytical tool to aid in gathering data and conducting analyses, not the least because available resources typically will not permit the labor-intensive efforts needed for the initial effort.

- **Contract Action Reporting System (J001).** This system records all Air Force purchasing office contract transactions of at least $25,000 on Form DD 350, the Individual Contract Action Report. The system contains a great deal of information, including contract characteristics, such as sole-source; funds committed to outstanding work orders, i.e., obligated dollars, and funds that have been de-obligated by a contract change or cancellation; the end item or weapon system for which funds were expended; the supplier; and the dominant category of good or service purchased in the transaction.
- **Acquisition and Due-in System (J041).** This AFMC acquisition system records post-award spares transactions with details on item quantities, costs, and transaction dates.[3] (These data are accessed through the J018R system.)
- **Contract Depot Maintenance and Cost System (G072D).** This Air Force Materiel Command Financial Management (AFMC/FM) legacy system provides financial and managerial information data for items under contract depot maintenance. It tracks repair requirements and provides funding information. It contains contract information aggregated by fiscal quarters on total quantities repaired by item, the unit sale prices to Air Force customers, and unit repair costs, but it does not record transaction-level data. Nor does the G072D system record individual contract action data, such as award dates.[4]
- **Automated Budget Compilation System (ABCS) (D075).** ABCS data are processed twice a year, in September and March, to calculate spare and repair requirements for budget justification purposes. (AFMC Logistics consolidates the spare and repair budgets into a single package for review and approval.) The system provides a detailed list of future item requirements, in-

---

[3] Post-award transactions include orders for repairs and spares that are made off of active contracts.

[4] Since October 2002 (after the period covered in these spend analyses), this information has been recorded in G072I, a mirror system of G072D, designed to hold information until a system to replace G072D becomes operational.

cluding quantities and costs (in current dollars) for the current fiscal year and subsequent three fiscal years. AFMC/FM uses the ABCS to develop separate buy and repair budgets for repairable items, i.e., items that can be repaired more economically than they can be replaced.

- **Item Manager Wholesale Requisition Process (D035A).** This system, designed for materiel management and customer support in an online, real-time, wholesale requisitioning process, is part of the Stock Control System encompassing requisition processing, inventory accounting, and returns management.

- **Bill of Materials (BOM) (D200F).** BOMs list material and components needed for manufacture, overhaul, or repair of an end item, assembly, or subassembly. BOMs are used for budget forecasts, workload plans and schedules, and projecting shop parts needs. The D200F system contains information on relationships among items and assemblies that fit into other items and assemblies, including information, for example, on items associated with the same weapon system or with assemblies of the same weapon system. This information is used in computing item requirements (D200A) and maintenance requirements (G005M). We used information from two types of BOMs in this study. First, we used an Automated or Actual BOM produced through a menu-driven system allowing for additions or changes to, or inquiries and deletions of, BOM records. Data from this source, according to the OC-ALC, reflect the actual shop experience of the items found on a weapon system. Second, we used Planning BOMs, based on engineering and configuration data.

- **Contracting Business Intelligence System (CBIS).** This AFMC contracting system is designed to be the clearinghouse for Air Force contract data. It provides a user-friendly interface for extracting information from legacy systems. CBIS contains contract information from DD350 and DD1057 data sources, along with information on requirements, solicitations, and accounting data. CBIS was in development in FY 2002 and was not available for the F100 PSM demonstration spend analysis.

- **Acquisition Method Code Screening System (J090A).** This system provides information on the ALC's assessment of market competitiveness for a spares item. It also provides a list of known qualified suppliers during a given period.

DLA manages most of the military services' common consumable items that are removed and replaced with new parts rather than repaired.[5] DLA data include data from the following sources:

- **Active Contract File (ACF).** In FY 2002, this file contained a nearly complete history for each DLA contract transaction, including contract number, total obligated dollars, commodity price, commodity quantity, and the National Stock Number (NSN), a part's identification number. To extract information on weapon system–related NSNs, DLA must identify the NSNs associated with Weapon System Designator Codes (WSDCs).
- **Requisition File.** The DLA requisition file records requests for items by date, NSN, quantity, and military service (i.e., the address of the requesting organization). Air Force market-share data come from this file.

DCMA data include those data from the Mechanization of Contract Administration Services (MOCAS) system. The DCMA and the Defense Finance and Accounting Service use the MOCAS system as a post-award contract administration disbursing system for most contracts monitored by the DCMA. The MOCAS system provides transaction-level details on scheduled and actual deliveries, and

---

[5] On November 9, 2005, the FY 2005 Base Realignment and Closure legislation became law (Miles, 2005). Among its recommendations was the relocation of procurement management of depot-level reparable spares from the services to DLA. This relocation must begin no later than FY 2007 and be completed by FY 2011 (Defense Base Closure and Realignment Commission, 2005). This report was in its final editing stage when these recommendations became law. Note that some of our observations regarding the benefit of including DLA parts with Air Force parts on contracts will occur as a consequence of this relocation. However, because the relocation of procurement management applies only to spares, the Air Force may still benefit from coordinating DLA's purchases of replacement parts and consumables with the Air Force's repair requirements.

automatic closure of contracts as prescribed by Federal Acquisition Regulations. Scheduled and actual deliveries are listed in two separate files that must be joined by transaction and date. NSN data were not included in the MOCAS system data that we received.

Other relevant data include the Data Universal Numbering System (DUNS) of Dun and Bradstreet, Inc., then purchased monthly by the Washington Headquarters Services, Statistical Information Analysis Division (formerly known as the Directorate for Information Operations and Reports) in the Office of the Secretary of Defense. The DUNS is a proprietary database of nine-digit numeric codes corresponding to firms and their facilities. These data provide information on relationships between parent and local firms (i.e., "parent/child relationships").

As noted, no single source of available data suffices to answer the questions raised in a spend analysis. In fact, in the case of the F100 spend analysis, not only was it necessary for data sources to be integrated for a comprehensive analysis, the data sources had to be integrated to answer nearly every question that a spend analysis seeks to answer (see Table 2.1 for a list of such questions and the data sources for answering those questions).

J001 data, for example, contain details on the contract and supplier for transactions worth at least $25,000, but little information on the actual goods and services purchased. This is because for each transaction, J001 permits the entry of only one Federal Supply Class (FSC) or Product and Service Code (PSC) and one weapon system.[6] The lack of detail on all individual goods and services purchased would limit use of this data source largely to strategic, high-level analysis and for contract and supplier-specific information.[7]

---

[6] FSC codes are similar to the North American Industry Classification System codes (albeit more finely grained and covering a narrower range) of industries producing goods ranging from clothing and food to ammunition and weapons. (PSCs also provide a finely grained classification of services purchased by the federal government. FSCs are used for goods and PSCs are used for services.)

[7] See Dixon et al. (2005), which includes an analysis of the data fidelity of DD350 FSC and PSC coding entries.

Table 2.1
Spend Analysis Questions and Available Data Sources for Answers to the Questions

Spend Analysis Question	Data Source
What was purchased?	J041, G072D, ACF
Who is buying and how much are they buying?	J001, J041, G072D, ACF
Who are our suppliers?	J001, J041, G072D, ACF
What are the characteristics of purchased goods and services?	
Weapon system, next-higher assembly	J001, ACF, D200F/BOM
Commodity or material type	J001, J041, G072D, ACF
Competitiveness	J090A, J001, ACF
What is the supplier's past delivery performance?	MOCAS
What supply strategies are required for the future, e.g., how do actual demands compare with projected ones?	ABCS/D075, D035A, DLA requisitions, J090A

J041 data contain, for all transactions and by item, detailed characteristics such as contract number, transaction order number, NSN, contract dollars, quantities, prices, award dates, and other characteristics, but they have no information on supplier or contract characteristics other than a supplier identification code and contract number. Matching J001 data on supplier characteristics and J041 data on item characteristics yielded more data by RAND on individual transactions than would otherwise be available, and, as discussed further in the next section, helped in data cleansing efforts.

Some Air Force data sets, e.g., J041 and G072D data, can be used to document what the Air Force has purchased but not what other logistics agencies, such as DLA, have purchased for the Air Force. Air Force and DLA data systems both have information on suppliers, but different Air Force data systems occasionally list different suppliers for the same transaction, necessitating the use of multiple data sources recording the same transactions to corroborate data accuracy, such as ensuring that the right supplier was identified for a transaction. In addition, the J001 data provide information on deobligated dollars. Air Force spare and repair and DLA contract data offer information on the characteristics of goods purchased for weapon systems, but D200F/BOM data are needed to show relationships among goods at lower levels of assembly, such as engine modules associated with engine types. MOCAS data can provide information on one aspect of past performance—on-time delivery—but these data need to be integrated with transaction and NSN-level data (e.g., contract number and purchase or delivery order number) to make them useful for developing supply strategies. Regarding actual and projected demands, the Air Force J041 and G072D databases provide information on actual NSN-level purchases, and the ABCS/D075 data provide information on future purchase needs; data from all three are needed to compare how well ABCS/D075 projects future Air Force purchase needs.

Data Cleansing and Validation

Like private-sector data systems, Air Force data systems contain inaccuracies, omissions, and ambiguities. Eliminating all such problems may not be feasible or necessary, but failure to address the most egregious problems can lead to grossly inaccurate conclusions.

When we first extracted F100 engine records from J001 data and ranked the top parent supplier companies, it appeared that the second most important supplier was General Electric, which manufactured the F110 engine, among others. This finding, however, turned out to be the result of a few erroneous weapon-system code entries for the purchase of new engine modules.[8] For some contracts, we found inconsistent information on contract and firm characteristics (e.g., whether a contract was competitive or sole source, whether a firm qualified as a small or disadvantaged business).

Some inconsistencies in contract data can arise not from errors but from the nature of a transaction. Basic ordering agreements, for example, can include both competitive and sole-source transactions

---

[8] Once the erroneous entries were identified, we removed these records from the F100 spend analysis.

and multiple suppliers. And some contracts may pertain to more than one weapon system. Nevertheless, the characteristics of most contracts remain fixed over time; thus, for most contracts, inconsistencies tend to indicate miscoded or otherwise erroneous information.[9] Such tendencies led us to develop a business rule (as recommended by the OC-ALC) to use the earliest transaction recorded for each contract, with data going back as far as 1994, to resolve any discrepancies appearing in subsequent transactions.[10]

There were also discrepancies among J001, J041, and G072D data in dollar amounts of transactions. In some cases, J001 transaction dollars exceeded values listed in J041 or G072D data. Because not all J001 transaction dollars are nominally related to F100 engine items (i.e., because such dollars may apply also to goods for other systems or for services unrelated to specific NSNs), these dollars could exceed the detailed contract data dollars. In such cases, we relied on the J041 or G072D values as shown in the data sets.

In other cases, transaction-level dollar values in J041 or G072D data exceeded those in J001 data. To test the relative accuracy of the differing dollar amounts in these cases, we identified transactions in 59 contracts from an FY 1999–2001 sample of transactions valued at $25,000 or more in both the J001 and J041 databases for which J041 transaction dollars for spare parts exceeded the J001 transaction dollars for spare parts. For 32 percent of the transactions we identified as having dollar differences between the two databases, OC-ALC staff compared the J041 and J001 records with original copies of the contract transactions found in actual ALC contract files. They discovered that in 78 percent of the cases, the J001 data dollar values were closer

---

[9] The characteristics of blanket purchase agreements, on the other hand, may differ by type of transaction (e.g., competitive or sole source, or supplier). Multiple suppliers are qualified to conduct the work and then bid on each purchase or delivery order.

[10] In successive spend analyses we conducted for other RAND studies, we analyzed the data with no modifications to make the analyses reproducible. Because most spend analyses are used internally by companies for making business decisions, many companies cleanse their data. Company representatives whom we interviewed said that the money they spend in cleansing their data is more than offset by the money they save in using more accurate, complete data for their processes and analyses.

to the values in the hard-copy contract file data than were the J041 data dollar values.[11] Further, according to the J041 database chief, de-obligations are entered as a lump sum for a contract number and are not allocated to NSNs.[12] Thus, J041 data do not include de-obligated dollars at the NSN or transaction level. This led us to rely on J001 data for all cases in which J041 transaction dollar values exceeded J001 transaction dollar values by more than $25,000, which was the minimum dollar threshold for J001 data in FYs 1999–2002. We adjusted or reconciled the J041 data to values that corresponded to those reported in the J001 system.

Altogether, contracts listed in both the J001 and J041 databases accounted for 4,856 transactions. Of these transactions, 42 percent required adjustment, representing 29 percent of the approximately $1.836 billion spent on these purchases. The actual adjustments amounted to 19 percent of the original dollars for the transactions in common between these databases and 3 percent of all contracts and data in the spend analysis.

There were also several large discrepancies, particularly in FY 1999 data, between the J001 and the G072D databases in dollar amounts of transactions. The dollar values recorded for repair services in the G072D databases were typically higher than those recorded in the J001 databases. OC-ALC personnel determined that these discrepancies were due to differences in the type of information recorded by various ALCs, with the San Antonio ALC, which handled F100 engine work until its closure in February 1999, recording requirements data and the Oklahoma City ALC recording obligations data.[13] Requirements are computed assuming no budget constraints, while purchases are subject to constraints, such as dollars and the existence of a contract or qualified source. As a result, requirement dollars can exceed purchase dollars.

---

[11] Email from Michael Yort, OC-ALC, May 29, 2002.

[12] Phone conversation with R. Scott Burk, OC-ALC, January 2003.

[13] The Air Force announced the results of the public-private competition for the San Antonio Air Logistics Center Propulsion Business Area workload, with Oklahoma City ALC winning the workload. See Cales (1999).

To adjust for cases in which G072D contract dollars exceed J001 contract dollars, we constructed a factor to be applied to the G072D data that would align these dollars to J001 levels. If G072D dollars were less than J001 dollars, no factor was applied.[14] The factor was the ratio of J001 dollars to G072D dollars by contract number. This factor was applied only when G072D dollars exceeded J001 dollars at the contract level.[15] It was applied to NSN quantities and dollars in the G072D data. In cases involving acquisition of both spare parts and repair services for the same contract, we applied the factor to both J041 and G072D data.

Altogether, contracts listed in both the J001 and G072D databases accounted for 797 transactions. Of these, 46 percent were adjusted, representing 21 percent of the approximately $760 million spent on these purchases.

We also discovered shortcomings in the BOM and the ABCS/D075 data for spend analyses. In cross-referencing detailed spend results to BOM data to link spend to engine modules and engine models, we found that less than half of the items in the spend analysis matched items in the BOMs. Therefore, the detailed spend data could not be linked to a specific engine model, such as the F100-PW-100 or F100-PW-229, nor to modules of the F100 engine, such as the core or augmentor. (Modules are one of the four engine units that can be separated for repair or overhaul and reassembled to produce a whole engine.) We were also unable to derive the precise relationship each NSN or item had with its next higher-level units. The BOM data we obtained from the OC-ALC contained indenture configuration information only on items related to engine type model series (such as the F100-PW-100), not modules (such as the core). A better match between the BOM and spend data may have permitted

---

[14] G072D data contain NSN-level dollars. J001 data contain all dollars on a contract, which can include dollars for purchases not associated with a specific item. Thus, for identical contracts, J001 dollar values can be expected to equal or exceed G072D dollar values.

[15] A sum of $25,000, the threshold at which a transaction appears in the J001 data, was also added to J001 contract dollars to allow for the possibility that not all transactions for a particular contract number were recorded in J001.

the development of spend analyses for particular modules, such as those experiencing high costs or poor performance.

Our spend analysis data also matched poorly with the ABCS/D075 data that we had hoped to use to estimate future spending requirements by supplier. Most NSNs in the F100 spend analyses had no corresponding match with NSNs in the ABCS data. This may be a result of repair NSNs being associated with "organic maintenance" (i.e., maintenance conducted at a government depot facility) rather than contract maintenance. Similarly, we found poor matches between J041 historical data and ABCS spares requirements data. The disparity between the ABCS and the J041 databases may be explained in part by the use of only a master stock number or preferred NSN in the ABCS. Alternatively, it may be that ALCs purchase only a subset of a determined requirement, or that actual demand differs significantly from ABCS forecasts.

Using DLA data required similar adjustments to overcome inconsistencies and other challenges. The biggest challenge in analyzing DLA spend data for a weapon system was determining the NSNs and dollars related to Air Force use. Most DLA-managed consumables are used across the services; therefore, to estimate Air Force spend for these consumables, it was necessary to estimate the proportion of purchases resulting from demand from the Air Force and those resulting from demand from other services.

To identify relevant DLA purchases for this analysis, OC-ALC first identified Weapon System Designator Codes for the F100 engine, and DLA then extracted records from its Active Contract File for the NSNs associated with those codes. DLA also provided Air Force market-share factors (defined as Air Force quantity demand divided by worldwide quantity demand) for each NSN by fiscal year. Many of the items that DLA purchases are common to systems across more than one military service. We therefore needed to develop a means of estimating the dollars spent on certain items to meet Air Force demands apart from the dollars spent by other services for the same items. We used these market-share factors to estimate Air Force spending for each NSN managed by DLA. To estimate DLA spending specifically for the F100 engine, we had hoped to develop an Air

Force engine market-share factor that applied to each Air Force engine. OC-ALC staff decided that the data necessary to develop Air Force engine market-share factors were not readily available. OC-ALC explored using D035K requisitions to estimate demand by location, given that most bases operate aircraft with similar engines. Knowing which base ordered an NSN might help to identify engine types associated with particular requisitions. This approach would work for most operation base locations, but not, at present, for an ALC, which also requisitions DLA-managed parts. OC-ALC repairs most Air Force engines and makes most of the requisitions for engine parts purchased by DLA. Unfortunately, Air Force requisitions to DLA do not identify the weapon systems requiring a common part. Thus, we were unable to estimate the relative demand by Air Force engine for common parts. Without requisition data by engine, we were not able to estimate purchases that DLA made just for the F100 engine. Spend for F100 items that were common to other engines was considered only as F100 spend, which overestimates DLA purchases for the F100 engine.

Challenges with data are to be expected when doing spend analyses, because spend analyses typically use data for purposes other than for those for which the data were originally intended. Indeed, spend analyses often raise more questions than they answer. Some of these questions were specifically listed earlier in this chapter; others are implied by the discussion in Chapter Five, particularly in the discussion on future research needs. Nevertheless, the available data offer several insights, presented in the next two chapters, into Air Force PSCM and how it might be improved both for the F100 engine system as a whole and for a single commodity, jet engine bearings, within the weapon system.

CHAPTER THREE

Spend Analysis Findings for the F100 Engine

One of the first steps that the F100 jet engine PSM demonstration team took was asking RAND to perform a spend analysis and to analyze key F100 suppliers and types of items purchased in FY 1999–2002. RAND examined the competitiveness of the contracts and the proportion of spare parts and repair services purchased from private-sector companies. Since the F100 jet engine PSM demonstration, AFMC reorganized its purchasing and supply-chain functions into Commodity Councils (teams organized to develop and execute tailored supply strategies for specific groups of goods and services) and constructed a Strategic Sourcing Analysis Tool to make spend analyses easier to perform.

The analyses described in this report are among the first of their kind performed for the Air Force, and although they represent a comprehensive analysis of total purchases made by the Air Force and the Defense Logistics Agency on behalf of the F100 engine program during a specified time period, we caution the reader against interpreting these analyses as a baseline for measuring PSCM savings and improvements in future years. Such a comparison would require additional information, such as total ownership costs as a function of engine usage rates, that would allow purchasing to be compared from year to year to track reductions in total ownership costs.

F100 Engine Modules

The F100 engine has five major modules that can be removed from the engine, serviced separately, and then reassembled: (1) inlet/fan, (2) core, (3) low-pressure turbine, (4) augmentor, and (5) gearbox (see Figure 3.1). The core module, also referred to as the "hot section" module, includes three major assemblies: the high-pressure compressor, combustor, and high-pressure turbine. The Air Force has four F100 engine models: the F100-PW-100, the F100-PW-200, the F100-PW-220/220e, and the F100-PW-229.

Who Is Purchasing Goods and Services for the F100 Engine?

In analyzing total F100 spend, we considered everything that the Air Force, DLA, and other military organizations spent on the F100 engine during FYs 1999–2002. We found that there are four principal

Figure 3.1
Modules of the F100-PW-220 Engine

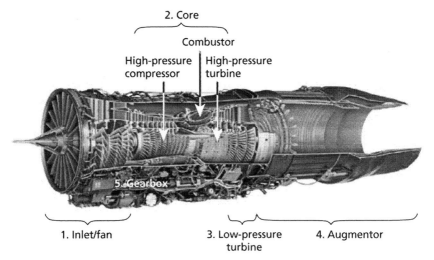

purchasers of F100 goods and services, and very little purchasing of these items by other organizations.

The first purchasing organization is the Oklahoma City ALC, which received the F100 workload from the San Antonio ALC after the Base Realignment and Closure (BRAC) Commission closed the San Antonio facility in 2001. OC-ALC is responsible for sustainment of the F100 engine. It purchases spare parts and repair services for reparable items and equipment modification services. It also purchases a small portion of the engine's consumable parts (i.e., parts that are replaced rather than repaired upon failure) for items deemed fracture- or safety-critical (i.e., items for which a fracture or other failure could lead to loss of a weapon system or human life). The Oklahoma City ALC and, previously, the San Antonio ALC, also purchase research, development, and technical evaluation (RDTE) services.

The second purchasing organization is the Aeronautical Systems Center (ASC) at Wright-Patterson Air Force Base, which purchases replacement engine modules for condemned modules considered to be beyond economical repair (i.e., modules that an expert has determined are cheaper to replace than to repair) and modification kits to modernize and retrofit the F100 engine to address technical and support issues. ASC also purchases RDTE, technical representative services, and aircraft demonstration and validation services required to ensure the capabilities of new equipment.

The third organization, DLA, purchases consumable items that are used by multiple military services and that are not fracture-critical or safety-critical. The Air Force also purchases from DLA gas turbine and jet engine components and engine bushings (which are similar, but not identical to, bearings).

Finally, a small portion of purchases for the F100 engine are made through contracts written by other military services and agencies, such as the General Services Administration (GSA). The GSA purchases commodities and services common to all government agencies (e.g., office supplies, travel services). Air Force purchases from the GSA during FYs 1999–2002 primarily were for data analyses (36 percent of Air Force F100 spend with the GSA) and engineering technical services (24 percent of Air Force F100 spend with the GSA). From Navy contracts, the Air Force purchased engine fuel system components and electrical boards and associated hardware.

What and How Much Are Organizations Purchasing?

In fiscal years 1999–2002, total annual F100 spend varied between $800 million and $1.1 billion (see Figure 3.2). The totals varied not just from year to year but also by purchasing organization and commodity.

In Figure 3.2, "Air Force items" include both spare parts and repair services that can be linked to an NSN. "Air Force other" includes those purchases that cannot be linked to an NSN; it primarily includes acquisition and testing of new F100 equipment and a small amount of engine sustainment expenditures for which no NSN-level details were available, because these purchases included goods or

Figure 3.2
Total F100 Engine Spend by Year, FYs 1999–2002

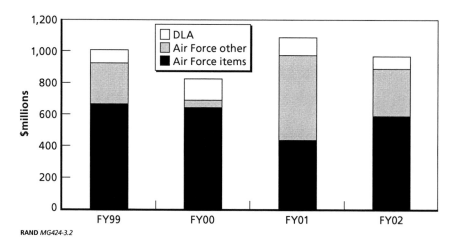

RAND MG424-3.2

DATA SOURCES: Air Force J001, J041, and G072D data; DLA Active Contract File data.
NOTE: Dollar amounts are adjusted to FY 2002 constant dollars.

services not identified with a particular NSN. DLA spend, as depicted in the figure, overestimates F100 spend because some DLA parts could also be used for other engines. Data to derive F100 application factors for those parts were unavailable.

Spend on Air Force F100 items was $670 million in FY 1999 before decreasing to $439 million in FY 2001 and then increasing to $593 million in FY 2002. This change in spend is attributable to spending patterns caused by the transition of the engine and overhaul depot maintenance workload to the OC-ALC after the closure of the San Antonio ALC. To minimize the risk of supply shortages during the transition, spending on repair contract services and other items was increased in FY 1999 and FY 2000, the two years prior to the transition, and then was sharply reduced in FY 2001, the year of the move to the San Antonio ALC.

The transition from the San Antonio ALC to the Oklahoma City ALC also affected other Air Force spending for acquisition and testing of F100 repair equipment. Spending narrowed from $256 million in FY 1999 to $48 million in FY 2000 before increasing to $540 million in FY 2001 when the new site for F100 work was established. Spending then decreased to $297 million in FY 2002.

By contrast, DLA spending has been relatively stable, averaging $102 million per year, but it, too, has varied—between $80 million and $133 million.

Across the four years of this analysis, most F100 spending was for ALC acquisition of reparable items and services, as shown in Table 3.1. The first row, "ALC items," represents ALC spend contained in the detailed NSN databases, J041 and G072D. All detailed NSN spend data were considered to be ALC spend. The second row, "ALC other," is the ALC spend in J001 not duplicated in the J041 and G072D databases. ALC spend in the J001 database can be identified by purchasing-office codes. The third row shows the Air Force's share of DLA spend for F100 items. The first three rows of data representing sustainment spend constituted more than three-fourths of F100 spend in the period of this study. ASC equipment purchases, shown in the fourth row of the table, accounted for less than one-fourth of F100 total spend.

Table 3.1
F100 Spend by Purchase Category and by Proportion of Spend on FSC 2840, FYs 1999–2002

Purchase Category	Average Annual Spend (constant FY 2002 $millions)	Percentage of F100 Spend	Percentage of Purchase Category Spend on FSC 2840
ALC items	597	61	88
ALC other	64	7	57
DLA	102	10	N/A
ASC	219	22	96
Total	982	100	N/A

N/A = data not available.

Within each F100 purchase category, the overwhelming share of spend is for goods in Federal Supply Class 2840—aircraft gas turbine and jet engines and components. The Air Force spends more on FSC 2840 than on any other single FSC, except fixed-wing aircraft (Moore et al., 2004).[1]

Altogether, 88 percent of item-related spend for the F100, or $525 million of the annual average item-related spend of $597 million, from FYs 1999–2002 is associated with FSC 2840. Figure 3.3 shows the subcategories that make up FSC 2840. The left side of the figure shows spending on subcategories of FSC 2840 for all four F100 engine models. The largest proportion of this spending (17 percent) was for the core module that is associated almost entirely with the F100-PW-229 engine and its Total System Support (TSS) contracts with Pratt & Whitney (PW), a division of United Technologies Corporation. Nearly all spending for the inlet/fan module (3 percent)

---

[1] Other FSCs for ALC items include the following (with percentage of average annual ALC item dollars): FSC 2915, engine fuel-system components (4 percent); FSC 3040, power transmission equipment (2 percent); and engine accessories (1 percent). Product and Service Codes for services not associated with a particular item or NSN include PSC J028, maintenance and repair of engines (17 percent); PSC K028, modification of equipment and engines (9 percent); PSC AC95, RDTE and engineering development (6 percent); FSC 4920, maintenance and repair shop specialized equipment (2 percent); and FSC 3110, unmounted anti-friction bearings (1 percent). The other PSC for ASC purchases is L015/L016, technical representative services (4 percent).

was also associated with the F100-PW-229. The Air Force considered these TSS contracts outside the purview of the demonstration, because they were large contracts that had been recently awarded. The right side of Figure 3.3 shows spending for subcategories of FSC 2840 for the other three engine models (the PW-100, PW-200, and PW-220/220e) whose contracts were mostly within the purview of the demonstration. These three engines are supported primarily by the Air Force with organic repair operations, many spares contracts and suppliers, and a few repair contracts and suppliers. After removing spending for the F100-PW-229, engine blades was the largest subcategory of goods within FSC 2840.

What Are the Potential Opportunities for Purchasing and Supply Chain Management Initiatives?

Spend analyses can provide insights into potential leverage that purchasers may gain with particular suppliers or groups of purchased

Figure 3.3
F100 Spend by FSC 2840 Items, FYs 1999–2002

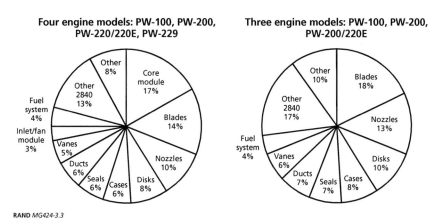

RAND MG424-3.3

DATA SOURCE: Air Force J001, J041, and G072D data.
NOTE: Data were adjusted to FY 2002 constant dollars.

goods and services. A spend analysis of F100-related purchases by the Air Force and DLA would indicate the relative leverage that each, as an F100 customer, has with its suppliers. Such an analysis can also indicate options that purchasers may wish to investigate further for gaining leverage and possible savings—e.g., leading purchasers might be able to offer expertise in coordinated buying efforts, or large numbers of contracts with the same suppliers or for similar items might be consolidated to reduce transaction costs.

Air Force purchases accounted for 85 percent of the dollars spent on F100-related contracts (see Figure 3.4). DLA accounted for far fewer dollars, but it had the overwhelming majority of contracts and had purchased items encompassing a far greater number of NSNs. Put another way, the Air Force manages fewer, more expensive F100 items and writes fewer, higher-value contracts for them. The Air Force purchases more expensive reparable items and relies on DLA for acquiring many cheaper consumable items. A large turnover of contract numbers from year to year suggests that DLA contracts have a short duration. This finding is consistent, as we later show, with the competitive nature of DLA contracts and the high proportion of contracts DLA awards to small businesses.[2]

Among Air Force contracts, most were for the sustainment of engines, including the manufacture of new replenishment spares, repairs, and engineering or technical support purchased through contracts identified with a particular NSN (see Figure 3.5). Remaining Air Force F100 contracts were for acquisitions. Although they are less numerous than sustainment contracts, acquisitions contracts were, on average, larger in value than those for sustainment.

Virtually all Air Force F100 purchases of items with NSNs were made by ALCs. The bulk of the F100 contracts, dollars, and NSNs were for spare parts (see Figure 3.6, which breaks out by spares and

---

[2] DLA was recognized as the DoD agency that awarded the greatest percentage of its prime contract dollars to small businesses in FY 2002; it awarded more than 38 percent of its contract dollars directly to small businesses, exceeding the government-wide goal of 23 percent. See Dearden (2003) and "Federal Procurement and Small Business Goals" (no date).

Figure 3.4
Air Force and DLA F100 Contracts, by Percentage of NSNs, Total Dollars, and Total Contracts, FYs 1999–2002

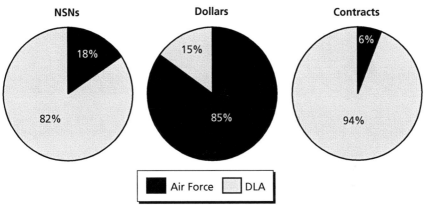

RAND MG424-3.4

DATA SOURCE: Air Force J041 and G072D data.
NOTES: Data were adjusted to FY 2002 constant dollars. F100-related contracts included non-F100-related transactions.

Figure 3.5
Percentage of Contracts and Dollars for Air Force F100 Sustainment and Acquisition NSNs, FYs 1999–2002

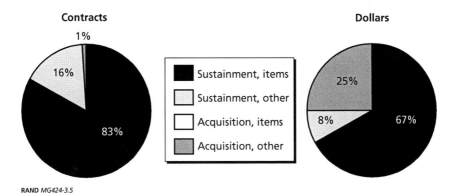

RAND MG424-3.5

DATA SOURCE: Air Force J001, J041, and G072D data.
NOTES: Data were averaged and are in constant FY 2002 dollars. Some contracts were counted more than once, because they contained both items and services not linked to NSNs or FSCs.

Figure 3.6
Percentage of Spares and Repairs in Air Logistics Center F100 Contracts for Sustainment, FYs 1999–2002

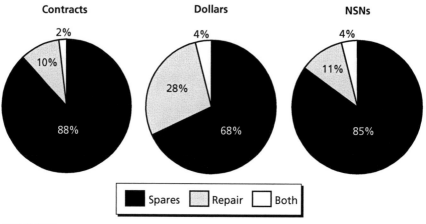

RAND MG424-3.6

DATA SOURCE: Air Force J041 and G072D data.
NOTES: Data are in constant FY 2002 dollars. Figure reflects only F100-related contract transactions with NSN-level data.

repairs the sustainment items shown in Figure 3.5). We discuss contracts for spare parts in more detail later in this chapter.

Turning to Air Force ALC purchases, most ALC F100 repair dollars (59 percent) were for a Pratt & Whitney TSS contract supporting the F100-PW-229. Much of the remainder (34 percent) was for depot maintenance "bridge" contracts used for transitioning the workload from the San Antonio ALC to the Oklahoma City ALC; those contracts are expiring over time and likely will not be replaced.[3] This means that only a small portion (7 percent) of ALC F100 purchases for repairs could be considered a prospective target for PSCM improvements, assuming future repair purchases remain proportionately the same.

---

[3] From FY 1999 to FY 2002, according to G072D data, the number of the bridge contracts decreased from 20 to seven, and their value decreased from $20.0 million to $3.3 million.

Similarly, a large proportion of Air Force F100 item purchases made through sole-source contracts present challenging prospective targets for PSCM improvements. Most Air Force F100 dollars, transactions, and NSNs were spent, conducted, or acquired through sole-source contracts, while the majority of Air Force F100 contracts were competitive (see Figure 3.7). Sole-source contracts were for more expensive items, averaging about $9 million, or nearly $700 per unit, while competitive F100 contracts averaged about $660,000, or nearly

Figure 3.7
Air Force F100 Spend for Sole-Source and Competitive NSN-Specific Purchases

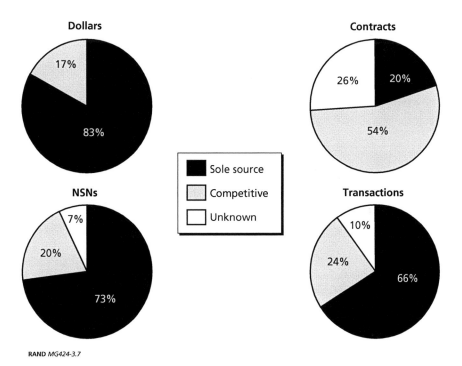

DATA SOURCE: Air Force J041 and G072D data.
NOTE: The percentages shown reflect primarily, but not exclusively, sustainment items.

$200 per unit.[4] Even the "competitive" contracts were almost uniformly limited to qualified sources, with contracts for only one NSN subject to full and open competition.[5] (These contracts could have included more than one NSN on them, but they showed activity or generated orders on only one.)

Sole-source contracts permit consolidation of requirements on the basis of particular providers, such as original equipment manufacturers (OEMs). These kinds of contracts often involve a greater number of items and are of higher dollar value than competitive contracts.

More than one in four contracts had no information on their competitiveness; such contracts, however, involved few dollars and only 7 percent of NSNs.[6] Nearly all of these contracts were manual purchase orders for less expensive, one-of-a-kind purchases. (Purchase orders are generally used when the transaction dollar value is less than $25,000; a manual purchase order is processed by hand.) We evaluated contract competitiveness from the J001 data, which did not record transactions less than $25,000 or many of the manual purchase orders.[7]

---

[4] J041 and G072D data indicate that competitive item unit prices were, on average, $195; sole-source item unit prices averaged $866; and items with an unknown competitiveness status had average unit prices of $220. Items on the TSS contract were more expensive; for example, non-TSS sole-source items cost, on average, $673. Engine modules on the TSS F110-PW-229 contract had average repair unit prices of $149,539. Altogether, on average, sole-source TSS items cost $434,357,865, while sole-source contracts for non-TSS items cost $7,011,575 over the four years we studied.

[5] "Qualified" sources are those that have a proven ability to perform work to Air Force specifications or that are distributors to whom original equipment manufacturers have agreed to sell parts. The use of qualified sources is common for engine parts because of their criticality. Full and open competition requires only proof to perform work to commercial standards, and thereby increases the number of potential suppliers that may compete for the contract.

[6] Information about contract competition came from the DD350 and J090 databases. The DD350 database reports transactions valued at or above $25,000, and the J090 contains the ALC's evaluation of the item's competitiveness.

[7] We prefer to use the J001 information on contract competitiveness, rather than the Acquisition Method Code (AMC), which classifies an item as competitive or sole source. AMC codes pertain to purchases of new spares and do not apply to contract repair. Also, a contract for an item evaluated by technical personnel as competitive may actually be sole source on the contract if only one company bid for the requirement. Use of only the AMC code can overestimate sole-source transactions for repair contracts.

Most DLA dollars for the F100 are spent on sole-source contracts (see Figure 3.8). Nevertheless, most DLA F100 contracts and transactions are from competitive sources, as are most of the individual NSNs the DLA acquires. The greater use that DLA makes of competitive sources for F100 items compared with sole-source contracts stems from its responsibility for purchasing consumable parts that are less complex to manufacture and cheaper to purchase. For some parts, DLA may choose between OEMs and distributors offering the same part.

Figure 3.8
DLA F100 Spend for Sole-Source and Competitive NSN-Specific Purchases

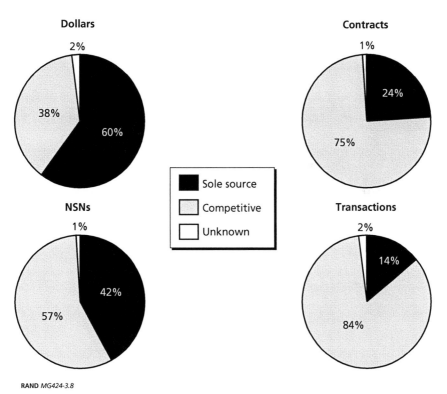

RAND *MG424-3.8*

DATA SOURCE: DLA Active Contract File data.
NOTE: Data are adjusted to constant FY 2002 dollars.

Although DLA spent less on the F100 than did the Air Force, it did so for far greater numbers of parts (NSNs) and through much larger numbers of contracts and suppliers (see Table 3.2). This suggests that its transaction costs were higher and its supply base larger that those of the Air Force.

Which Companies Might Help with Purchasing and Supply Chain Management Initiatives?

Spend analyses can identify key suppliers that the Air Force may want to target strategically for PSCM initiatives. The F100 spend analyses show that many of the companies that receive much of the Air Force's F100 spare parts and repair dollars are also the companies that receive much of the DLA's spare parts dollars. Market leverage can be improved if DLA works with the Air Force to develop supply strategies for items they both purchase from the same key suppliers in cases in which the Air Force's total business with a supplier exceeds that of DLA.[8]

The leading recipient of F100 dollars for both the Air Force and DLA was United Technologies Corporation (UTC), the parent company of Pratt & Whitney, the prime contractor for the F100 jet

Table 3.2
Characteristics of F100 Purchases and Supply Base for Air Force and DLA

	Air Force (Number)	DLA (Number)
NSNs	1,718	9,841
Supplier ID codes	157	1,858
Parent companies	142	1,695
Contracts	1,099	16,481

---

[8] Recent BRAC statutes call for moving purchases of all item spares to DLA. The Air Force's leverage with OEMs will continue to exceed that of DLA even after this move if acquisition purchases of new development programs and weapon systems are considered, because they exceed the total value of reparable and consumable spares purchases.

engine. The fact that UTC was the leading recipient was not surprising, particularly given the Pratt & Whitney TSS contract, but the relative magnitude of the spend with UTC on F100 engine support, which previously had not been thoroughly documented, was surprising. Other top F100 suppliers common to the Air Force and DLA include Honeywell International, Smiths Group (Tri-Industries), and General Electric (Unison Industries). At the same time, some of the top F100 suppliers for the Air Force and DLA differ, and the proportion of F100 dollars each spends with the prime contractor also differs.

Air Force spending for the F100 engine was concentrated in only a handful of suppliers. From FY 1999 to FY 2002, the Air Force spent, on average, $484 million per year for F100 items from UTC, an amount representing 82 percent of its $597 million per year for such items (see Table 3.3).[9] The top 15 Air Force suppliers represented 98 percent of total Air Force F100 dollars, and the top three suppliers—UTC, Dynamic Gunver Technologies, and Honeywell International—accounted for almost 90 percent of total Air Force F100 spend. These three parent companies provided more than 70 percent of the F100 NSNs the Air Force purchased and had 31 percent of the average annual number of F100 contracts. UTC and Honeywell also conducted the largest amount of sole-source business with the Air Force. (Most F100 dollars spent on General Electric [Unison Industries] items also were for sole-source items.) The remaining 10 percent of F100 spares and repair item spending, about $58 million annually, was spread among 139 other parent corporations holding 69 percent of the contracts and selling items for nearly 30 percent of the F100 NSNs.

---

[9] Some parent companies listed in Table 3.3 include, in parentheses, the names of units with which the Air Force conducts F100 business. For example, Chromalloy Gas Turbines is listed in parentheses with Sequa Corporation. Aggregating spend to the parent company rather than to separate divisions provides a more complete, strategic view of the business that the Air Force conducts with a supplier, which can be particularly useful when pursuing PSCM initiatives with multiple units of the same parent corporation. Such aggregation can also affect the relative ranking of suppliers.

Table 3.3
Air Force's Average Annual F100 NSN Spending, by Parent Company, FYs 1999–2002

Rank	Parent Company	Dollars (millions)	Percentage of NSN Spending	Percentage of Dollars from Sole-Source Contracts	Number of Contracts	Number of NSNs
1	United Technologies Corp.	483.9	82.5	94	64	546
2	Dynamic Gunver Technologies	23.9	4.1	5	22	18
3	Honeywell International	17.8	3.0	75	17	53
4	Sequa Corp (Chromalloy Gas Turbine)	12.4	2.1	38	7	15
5	Woodward Governor Company	10.0	1.7	38	9	14
6	Smiths Group (Tri-Industries)	7.4	1.3	4	8	9
7	SNECMA (Techspace Aero)	5.0	0.9	28	4	5
8	AAR Corp	3.4	0.6	14	8	11
9	Wood Group Fuel Systems	3.0	0.5	0	3	9
10	Goodrich (Rosemount Aero)	2.6	0.4	18	2	2
11	GE (Unison Industries)	1.8	0.3	53	12	12
12	Praxair, Inc.	1.7	0.3	0	3	8
13	Ferrotherm Corporation	1.4	0.2	14	4	4
14	Networks Electronic Corp.	1.1	0.2	10	2	1
15	Heroux-Devtek, Inc.	0.9	0.2	0	1	1
	Other	10.5	1.8	12	172	127
	Total	586.8	100.0	83	334	768

DATA SOURCE: Air Force J041 and G072 data.
NOTE: Data are averaged for FY 1999 to FY 2002 and adjusted to constant FY 2002 dollars.

By ranking F100 NSN spending by supplier, analysts can learn where to direct their efforts to improve PSCM. In particular, the concentration of Air Force F100 spend among top producers suggests that the Air Force may have some opportunities to improve its PSCM processes with those suppliers. For example, it might consolidate the number of contracts with its top suppliers, thereby reducing total transaction costs.

DLA's purchasing patterns for F100 items differed from those of the Air Force (see Table 3.4). For one, its annual average F100 spend of $102 million is about one-sixth that of the Air Force. Its spending was also much more widely distributed across a larger supply base. DLA had a greater number of F100 suppliers, including more small business suppliers, and wrote more contracts, particularly those of smaller value. Whereas the top 15 Air Force suppliers represented 98 percent of total Air Force F100 dollars, the top 15 DLA suppliers made up 63 percent of total DLA F100 dollars. While, as with the Air Force, DLA's F100 spending with UTC, the prime contractor, was more than that for any other supplier, it represented only 31 percent of the overall DLA F100 spend, compared with 82 percent for the Air Force. The difference in total dollars was even more striking; on average, the DLA F100 spend for UTC was $31 million, compared with $484 million for the Air Force.

Interestingly, the top DLA suppliers have a higher proportion of sole-source contracts than do the top Air Force suppliers, but, because DLA spending was distributed more widely, such supplier contracts constitute a small percentage of the DLA spend. There are several possible reasons for this finding. One reason may be the ability of small and disadvantaged businesses to compete for and win DLA contracts for parts that are true commodities, such as bolts and washers, or for contracts for parts that are less profitable and less attractive to large businesses. The data indicate that despite the Air Force's effort to open part of its F100 contracting requirement to competition, most of the total F100 contract dollar value went to sole-source contracts and suppliers. (As stated earlier, nearly all Air Force F100 spending with UTC, which received 82 percent of Air Force spending, was through sole-source contracts.)

Table 3.4
DLA's Average Annual F100 NSN Spending, by Parent Company, FYs 1999–2002

Rank	Parent Company	Dollars (millions)	Percentage of NSN Spending	Percentage of Dollars from Sole-Source Contracts	Number of NSNs	Number of Contracts
1	United Technologies Corp.	31.2	30.5	98.9	503	64
2	Honeywell International	11.5	11.2	87.5	319	73
3	GE (Unison Industries)	5.7	5.5	84.7	102	102
4	Whittaker Controls	3.4	3.3	86.2	57	39
5	New Hampshire Ball Bearings[a]	2.2	2.1	30.7	11	13
6	E.I. DuPont de Nemours	1.4	1.4	16.0	7	9
7	SPS Technologies	1.4	1.3	3.0	45	51
8	Smiths Group (Tri-Industries)	1.1	1.1	22.4	8	8
9	Marvin Engineering Co.	1.1	1.1	100.0	1	2
10	UMECO PLC	1.0	1.0	2.8	51	38
11	FAG Bearings Ltd.[a]	1.0	0.9	7.3	6	7
12	Eaton Corporation	0.9	0.9	74.9	24	24
13	Canadian Commercial Corp. (CCC)[a]	0.9	0.8	0.0	3	3
14	Wesco Aircraft	0.8	0.8	0.6	51	74
15	TPG N.V.	0.8	0.8	0.0	2	2
	Other	38.1	37.2	25.7	3,678	4,005
	Total	102.3	100.0	60.1	4,710	4,516

DATA SOURCE: DLA Active Contract File data.
NOTE: Data are averaged for FY 1999 to FY 2002 and adjusted to constant FY 2002 dollars.
[a] Original engine bearings manufacturers; FAG Bearings also sells through CCC.

To drive down total costs, both the Air Force and the Defense Logistics Agency may need to consider ways to shrink the total number of suppliers, especially among suppliers with whom they spend relatively few dollars and multiple suppliers that provide the same or very similar items. Among other benefits, such contraction of the supply base and the attendant contracts would give personnel time to become more familiar with the industries with which they work, including best practices in those industries, and more time to work with suppliers to implement PSCM initiatives (Moore et al., 2004). However, a competitive supply base would need to be preserved and socioeconomic goals would still need to be met.

Another way that the Air Force and DLA could improve their PSM practices for the F100 is by working collaboratively to develop supplier relationships and collectively leveraging through the agency with the greatest spend at a particular company, which the Air Force tried to do under the F100 demonstration.[10] Examining combined Air Force and DLA spending can help to identify the best areas for potential improvement through collaboration or consolidation. Among the top 15 F100 suppliers of items to the Air Force and DLA combined, 12 sell to both organizations, but some of these companies sell only to the Air Force, and only four are in the top 15 for both the Air Force and DLA (see Table 3.5).[11] Among the 12 companies that sell to both organizations, the Air Force has more F100 purchases, and presumably greater leverage, with eight suppliers, while DLA has more F100 purchases with four suppliers.

Unison Industries, Whittaker Controls, New Hampshire Ball Bearings, and Du Pont sold more F100 items to DLA than to the Air Force. All other corporations listed in Table 3.5 sold more F100

---

[10] Even with the move of depot-level reparable spares purchasing to DLA, DLA will still need to work with the Air Force on developing contracts with OEMs for complex, expensive items.

[11] SNECMA, Wood Group Fuel Systems, and Praxair sold F100 items only to the Air Force. Though listed as selling, on average, 100 percent of its F100 items to the Air Force, AAR Corp. sold a small amount of items to DLA between FY 1999 and FY 2002, ranking 849th among DLA F100 suppliers.

Table 3.5
Average Annual Combined Air Force and DLA F100 NSN Spending, by Parent Company, FYs 1999–2002

Combined Rank	Air Force Rank	DLA Rank	Parent Company	Dollars (millions)	Percentage of Dollars Spent by Air Force	Percentage of Total Dollars Spent by Air Force and DLA	Percentage of Dollars from Sole-Source Contracts	Number of Contracts	Number of NSNs
1	1	1	United Technologies Corp.	515.1	94	74.7	94.8	125	1,049
2	3	2	Honeywell International	29.3	61	4.2	80	90	372
3	2	46	Dynamic Gunver Technologies	24.2	99	3.5	5.1	28	23
4	4	52	Sequa Corp. (Chromalloy Gas Turbine)	12.6	98	1.8	37	13	18
5	5	171	Woodward Governor Company	10.1	99	1.5	38.7	13	18
6	6	8	Smiths Group (Tri-Industries)	8.5	86	1.2	6.1	15	17
7	11	3	GE (Unison Industries)	7.5	24	1.1	77.1	114	116
8	7	—	SNECMA (Techspace Aero)	5.0	100	0.7	27.5	4	5
9	44	4	Whittaker Controls	3.4	2	0.5	85.9	40	58
10	8	849	AAR Corp.	3.4	100	0.5	13.9	11	13
11	10	21	Goodrich (Rosemount Aero)	3.2	82	0.5	17.1	11	7
12	9	—	Wood Group Fuel Systems	3.0	100	0.4	0	3	9
13	17	5	New Hampshire Ball Bearings	2.9	26	0.4	22.8	15	13
14	12	—	Praxair Inc.	1.7	100	0.3	0	3	8
15	27	6	E.I. DuPont de Nemours	1.7	18	0.2	13.3	11	8
			Other	57.5	22	8.3	22.7	4,352	3,915
			Total	689.1	85	100.0	79.4	4,846	5,284

items to the Air Force than to DLA. A spend analysis conducted with respect to suppliers, rather than with respect to a particular category of goods and services (as was done here), would be needed to determine whether the Air Force or DLA had greater overall leverage with a particular supplier. Such a strategic approach would indicate where leverage could be applied—i.e., where consolidation of requirements may bring the most benefits. If the Air Force is a large customer of a supplier, and a group of goods or services is a large component of that business, then the Air Force has more opportunity to leverage its business with that supplier for greater gains in performance and reductions in cost. Alternatively, if the analysis shows that the Air Force is a minor customer, and the goods and services constitute a small portion of the overall business with the supplier, then the Air Force may wish to work with other major stakeholders in the Department of Defense to achieve performance and cost improvements, or to allocate fewer resources to improvements in cost and performance from suppliers with whom it has little leverage. Such analyses, however, were beyond the scope this study.

The data we studied indicate that the Air Force and DLA may realize strategic goals and benefits, by developing collaborative relationships led by the organization with the greatest amount of purchases and leverage over a particular supplier. Suppliers can also benefit through consolidation of the business they conduct with the Air Force and DLA, particularly in business planning and investment decisions.

Extending Analysis to a Specific Commodity

When the OC-ALC team reported the results of the initial F100 spend analysis, USAF/A4I, a sponsor of the F100 demonstration and of this study, asked the OC-ALC team to select an appropriate group of items to be the basis for writing a contract featuring PSCM improvements, such as development of a strategic supply strategy, market research, and analysis of contractors' past performance. OC-ALC's choices for PSCM improvements were somewhat limited. At the time, the OC-ALC was completing a corporate contract (a contract consolidating all the business a customer does with a supplier) for Pratt & Whitney parts and participating in a collaborative effort with DLA to form a strategic supplier alliance with Honeywell International. Existing F100-PW-229 TSS and "bridge" contracts were also not available for consideration as a PSM improvement target; the TSS contract was for whole engine support and was considered beyond the scope of the F100 PSM demonstration, and the bridge contracts were temporary and unlikely to be renewed after they expired. Among the remaining items not barred from the demonstration's consideration, OC-ALC chose jet engine bearings as a commodity group for PSCM improvements. Engine bearings are an integral part of engine repair, they are NSNs that apply to all engines, and they involve significant purchase dollars for both the Air Force and DLA.

RAND analyzed Air Force and DLA acquisition of jet engine bearings by using data that had been collected by OC-ALC for its own analyses. We present the results of the RAND analysis in the next chapter.

CHAPTER FOUR

Spend Analysis for Jet Engine Bearings

The OC-ALC had, as we discussed in the previous chapter, some limitations in its choices for PSCM improvements, which led it to focus on jet engine bearings for all Air Force engines. Even without the restrictions, there are several reasons for focusing on jet engine bearings as a potential target for PSCM initiatives. Past supply-chain and availability problems with this group of items had adversely affected readiness. More generally, there were long production lead times, diminishing manufacturing supply sources, and limited repair capability for F100 engine bearings (U.S. Air Force and Oklahoma City Air Logistics Center, 2002). Air Force customers also spend a comparatively large amount on bearings, which are key to parts rotations in engines, wear out under pressure use, and are among the more technically complex consumable items used in jet engines.

Per the assessment of OC-ALC engine experts, this spend analysis included several bearings from FSC 2840, such as bearings mounted on the engine case; all FSC 3110 bearings; unmounted anti-friction bearings; and all FSC 3130 mounted bearings. Jet engine bearings are consumable items that typically are replaced with new items when they fail. As a result, all the DLA and nearly all the Air Force jet engine bearings purchases in FYs 1999–2002, the period studied, were for spare bearings, although the Air Force also spent a small amount on repair services, primarily for bearings inspection and refurbishment.

How Are Jet Engine Bearings Purchased?

In FYs 1999–2002, the Air Force and DLA spent between $25 million and $33 million per year on jet engine bearings (see Figure 4.1). Unlike many F100 items, DLA purchases of this item far exceeded those by the Air Force. DLA spent from $16 million to $22 million annually, or an average of about $18.5 million annually, on spare bearings, whereas the Air Force spent an annual average of $8.7 million on spare bearings and an average of $1.6 million annually on repair services for bearings. Air Force spend for bearings decreased from FY 1999 to FY 2001 and increased in FY 2002.

During this period, DLA purchases of spares accounted for 59 percent of all dollars spent on jet engine bearings and 81 percent of all items purchased (see Figure 4.2). DLA purchased an annual average of 206,659 bearings, while the Air Force purchased an annual average of 12,621 bearings. Because the Air Force is responsible for more fracture- or safety-critical bearings than is DLA, it purchases

Figure 4.1
Total Jet Engine Bearings Spend by Year, FYs 1999–2002

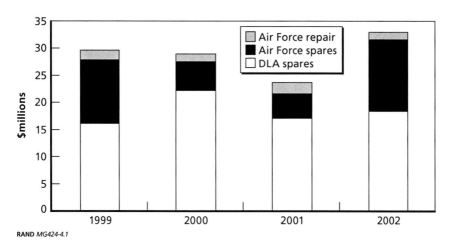

RAND *MG424-4.1*

DATA SOURCE: J041 and G072D NSN-level data on bearings-related contract transactions, FYs 1999–2002.
NOTE: Dollar amounts are adjusted to FY 2002 constant dollars.

Figure 4.2
Total Jet Engine Bearing Purchases, by Percentage of Dollars Spent and Items Purchased by the Air Force and DLA, FYs 1999–2002

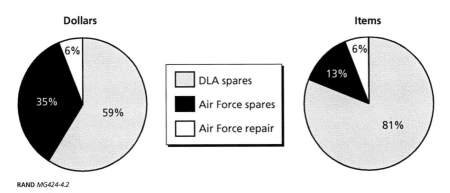

RAND *MG424-4.2*

DATA SOURCE: Air Force J041 and G072D and DLA ACF data.
NOTE: Dollar amounts are adjusted to FY 2002 constant dollars.

more-expensive bearings. The average price for an engine bearing that the Air Force purchased was $691, compared with an average price of $70 for an engine bearing purchased by DLA.

An analysis of detailed item-spend data for bearings shows that the large majority of Air Force contracts and dollars for bearings were for spares rather than repairs, and that more than three of four NSNs that the Air Force required for bearings were for spare items (see Figure 4.3).

Available detailed item-level data indicate that the Air Force spend on bearings was nearly split between competitive and sole-source providers, while DLA's spend was more competitive (see Figure 4.4). Regardless of supplier source, Air Force contracts had greater values than DLA contracts. In FYs 1999–2002, the average value of an Air Force contract with a competitive source for bearings was $293,624 (in FY 2002 dollars), while the average value with a sole source was $529,151. The average value of a DLA contract with a competitive source was $58,431, while the average value with a sole source was $65,634.

Figure 4.3
Percentage of Spares and Repairs for Cumulative Air Force Contract, Dollar, and NSN Bearings Spend, FYs 1999–2002

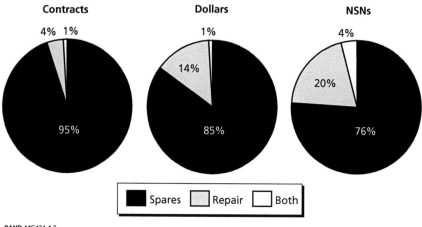

RAND MG424-4.3

DATA SOURCE: J041 and G072D NSN-level data on all bearings-related contract transactions, FYs 1999–2002.
NOTE: Dollar amounts are adjusted to FY 2002 constant dollars.

Figure 4.4
Air Force and DLA Spend Competitiveness for Jet Engine Bearings, by Percentage of Dollars Spent with Competitive and Sole-Source Providers, FYs 1999–2002

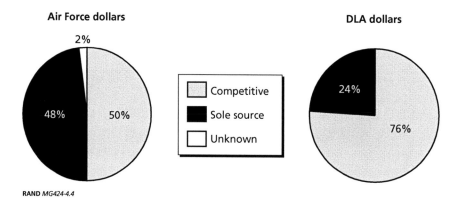

RAND MG424-4.4

Similarly, DLA acquired a greater percentage of bearings NSNs from competitive sources than did the Air Force (see Figure 4.5).[1] Most of the contracts for these purchases had relatively short durations, although the sole-source contracts tended to have longer duration. About 3 percent of the DLA contracts for NSNs with competitive sources had durations of at least four years, compared with 11 percent of DLA contracts of at least four years in length for NSNs from sole sources.[2] For the Air Force, 9 percent of contracts for NSNs with competitive sources and 39 percent for NSNs with sole sources had durations of at least four years.

The differing levels of competition in the markets from which the Air Force and DLA purchase bearings indicate differing challenges for organizations wishing to maintain a competitive supply base while also enjoying the benefits of having strategic relationships with suppliers. The Air Force has traditionally tried to reduce costs

Figure 4.5
Air Force and DLA Competitiveness for Jet Engine Bearings, by Percentage of NSNs from Competitive and Sole-Source Suppliers, FYs 1999–2002

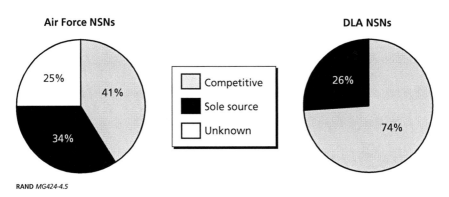

RAND *MG424-4.5*

---

[1] The large proportion of Air Force NSNs for which competitive source data were unknown was typically associated with manual purchase orders averaging about $4,700. Information on contract competitiveness is based on DD350 data, which omit many contract transactions of less than $25,000.

[2] Some contracts may have had longer durations. Because we analyzed only four years of data, we could not determine the exact duration of the contracts.

through competition whenever possible, seeking in particular to reduce unit price. PSCM best practices seek to reduce total costs by developing better relationships with key suppliers and improving supply chain processes and product quality. Many of these best practices are implemented outside of specified contract language but ultimately lead to improved contracts with key suppliers. Significant improvements in the relationships between suppliers and customers are not possible with a large supply base—creating conflict between efforts to reduce costs through increased competition and efforts to improve supply chain processes and supplier relationships more generally.

Who Are the Leading Suppliers?

Spend on bearings for both the Air Force and for DLA was concentrated among the top suppliers. For the Air Force, 86 percent of bearings spend for both spares and repairs went to the top five suppliers (see Table 4.1). Only two of the five, AB SKF and MPB Corporation (Timken),[3] were OEMs. Other OEMs for bearings were FAG Bearings, New Hampshire Ball Bearings, and Torrington Industries. (Torrington had not been an Air Force supplier before its acquisition in 2003 by Timken,[4] although since then it has been part of a corporation that is an Air Force supplier.) Overall, the Air Force spent 60 percent of its bearings dollars with OEMs or with the Canadian Commercial Corporation, a contract clearinghouse for Canadian companies and intermediary party that sells bearings for FAG Bearings.

United Technologies Corporation, the single largest supplier of bearings to the Air Force by dollars, manufactures parts of the engine in which bearings are housed, not the bearings themselves. Many of

---

[3] Timken acquired MPB Corporation in May 1990. See the Securities and Exchange Commission Web site, http://www.sec.gov, Timken annual report submitted March 31, 1994.

[4] Timken acquired Torrington Industries in February 2003. See the Securities and Exchange Commission Web site, Timken annual report submitted March 3, 2004.

Table 4.1
Average Annual Air Force Jet Engine Bearings Spending, by Parent Company, FY 1999–2002

Rank	Parent Company	$Millions	Percentage of All Air Force Bearings Spend Dollars	Percentage of Dollars Spent with Sole-Source Contracts	Number of Contracts	Number of NSNs[a]
1	United Technologies Corp.	3.8	37	96	4	16
2	AB SKF (MRC Bearings)[b]	1.6	16	13	10	9
3	MPB Corp. (Timken)[b]	1.5	15	3	5	5
4	Canadian Commercial Corp.	1.2	12	0	2	1
5	Honeywell International	0.7	7	56	3	4
6	Bearing Inspection (Timken)	0.5	4	69	2	17
7	FAG Bearings (Precision Bearing)[b]	0.4	4	30	3	2
8	Lockheed Martin	0.2	2	0	1	5
9	General Electric	0.1	1	53	2	3
10	New Hampshire Ball Bearings[b]	0.1	1	52	1	1
11	Alamo Aircraft	< 0.1	< 1	73	4	3
12	Eagle Industries	< 0.1	< 1	0	4	1
13	Galaxy Aircraft Parts	< 0.1	< 1	0	5	3
14	Westfieldgage Co.	< 0.1	< 1	100	1	1
15	Amjet Aerospace	< 0.1	< 1	0	1	1
	Other	2.3	<1	23	6	4
	Total]	10.3	100	48?	53	75

DATA SOURCE: J041 and G072D data.
NOTES: Data are averaged and adjusted to FY 2002 constant dollars. Some parent companies listed in this and the following tables include, in parentheses, the names of units with which the Air Force conducts F100 business.
[a]NSNs may be supplied by more than one company.
[b]Original equipment manufacturers.

the bearings that UTC sold to the Air Force were those it has incorporated into UTC items. Some of the high-value bearings-related items that UTC sold to the Air Force include seal supports, air seals, and bearings housings. Similarly, many of the bearings that Honeywell sold to the Air Force were those that Honeywell installed on engine parts that it manufactured.

Other distributors who buy bearings from OEMs were able to become qualified sources of supply and compete directly with the manufacturers. At a meeting of members of the RAND study team and the OC-ALC demonstration team, OC-ALC team members said that some OEMs have some ambivalence about conducting business directly with the Air Force. Being unfamiliar with government contracting procedures, the OEMs have instead sought to sell to distributors who sell bearings to the government. The significant barriers to entry in manufacturing in this industry mitigate the risk of OEMs facing "reverse engineering" of their products by potential competitors.

OEMs do not necessarily have sole-source contracts with the Air Force. For example, most of the dollars spent with Timken were for items that are available from other sources. This suggests that OEMs compete with distributors that are qualified sources of supply. Most of the competitive contracts for FYs 1999–2002 involved distributors of spares, not alternative sources of supply. Such distributors compete with OEMs through lower transaction (i.e., indirect) costs; OEMs have long argued that they cannot afford to tailor their business processes to military needs, unless there is sufficient business to warrant the extra transaction costs for doing so. For companies that do not have separate business units dedicated to defense customers and have most of their business with commercial customers, the cost of selling directly to DoD and possibly tailoring their internal processes and data systems to DoD's needs can be expensive.

For DLA, 70 percent of the bearings spend went to the top six suppliers (see Table 4.2). OEMs ranked higher among DLA bearings suppliers than among Air Force bearings suppliers. All five OEMs, plus Canadian Commercial Corporation, which sells FAG Bearings, ranked among the top seven suppliers of bearings to DLA and received 62 percent of DLA's bearings spend.

Combining data on individual suppliers shows that DLA had more leverage with key bearings manufacturers than did the Air Force (see Table 4.3). Among bearings OEMs, the Air Force had leverage similar to DLA's with only Timken. Its leverage could be greater with

Table 4.2
Average Annual DLA Jet Engine Bearings Spending, by Parent Company,
FYs 1999–2002

Rank	Parent Company	$Millions	Percentage of All DLA Jet-Engine Bearings Spend Dollars	Percentage of Dollars Spent with Sole-Source Contracts	Number of Contracts	Number of NSNs
1	AB SKF (MRC Bearings)[a]	4.4	24	19	31	27
2	Honeywell International	2.1	11	75	3	13
3	Canadian Commercial Corp.	2.0	11	0	5	5
4	New Hampshire Ball Bearings[a]	1.7	9	9	21	18
5	FAG Bearings (Precision Bearing)[a]	1.6	8	39	16	12
6	MPB Corp. (Timken)[a]	1.1	6	5	20	19
7	Torrington (acquired by Timken)[a]	0.7	4	4	15	13
8	Jamaica Bearings	0.6	3	7	27	29
9	United Technologies Corp.	0.4	2	14	5	10
10	Accurate Bushing	0.4	2	0	11	8
11	Kaman Industrial Technologies	0.3	2	0	1	1
12	White Engineering Surfaces	0.3	2	0	1	1
13	Thomson Industries	0.3	2	100	2	1
14	Roller Bearing Holding Co.	0.2	1	0	7	6
15	General Electric	0.2	1	36	2	8
	Other	2.1	11	38	86	54
	Total[b]	18.5	100	25	253	225

DATA SOURCE: J041 and GO72D data.
NOTE: Data are averaged and adjusted to FY 2002 constant dollars.
[a]Original equipment manufacturers.
[b]Totals may not be exact due to rounding.

UTC with whom it spends more dollars than it does with Timken, primarily through sole-source contracts.

DLA spent more with several bearings suppliers than did the Air Force, but the Air Force or another service had a higher overall average annual spend with many bearings suppliers (see Table 4.4). DLA spent more than the Air Force on jet engine bearings from the Canadian Commercial Corporation ($2 million versus $1.2 million) and

Table 4.3
Average Annual Combined Air Force and DLA Jet Engine Bearings Spending, by Parent Company, FYs 1999–2002

Combined Rank	Air Force Rank	DLA Rank	Parent Company	$Millions	Percentage of Total Bearings Spend Dollars	Percentage of Air Force Bearings Spend Dollars	Percentage of Dollars Spent with Sole-Source Contracts	Number of Contracts	Number of NSNs
1	2	1	AB SKF (MRC Bearings)[a]	6.0	22	27	17	41	36
2	1	9	United Technologies Corp.	4.2	18	89	89	9	26
3	4	3	Canadian Commercial Corp.	3.2	12	38	0	7	6
4	5	2	Honeywell International	2.8	10	26	69	6	17
5	3	6	MPB Corp. (Timken)	2.7	11	56	4	25	24
6	7	5	FAG Bearings (Precision Bearing)[a]	1.9	7	20	37	19	14
7	10	4	New Hampshire Ball Bearings[a]	1.8	6	6	12	22	19
8	—	7	Torrington (Timken)[a]	0.7	2	0	4	15	13
9	—	8	Jamaica Bearings	0.6	2	0	7	27	29
10	6	47	Bearing Inspection (Timken)	0.5	2	90	69	2	18
11	—	10	Accurate Bushing	0.4	1	0	0	11	8
12	8	16	Lockheed Martin	0.4	2	59	2	1	6
13	—	11	Kaman Industrial Technologies	0.3	1	0	0	1	1
14	—	12	White Engineering Surfaces	0.3	1	0	0	1	1
15	9	15	General Electric	0.3	1	45	44	4	11

[a] Original equipment manufacturer.
NOTE: — = not applicable.

Honeywell ($2.1 million versus $1.7 million) (see Table 4.3), but the Air Force had higher total expenditures with these suppliers than did DLA, and other services spent even more with those suppliers than did the Air Force (see Table 4.4). This finding suggests that DLA's leverage with these companies may be somewhat limited when compared with the strategic business base of the Air Force and other military services.

Table 4.4
Total FY 2002 Spend with Top Bearings Suppliers by Air Force, DLA, and Other Services

Combined Air Force and DLA Rank	Parent Company	Annual Expenditures ($millions)		
		Air Force	DLA	Other Services
1	AB SKF (MRC Bearings)[a]	2	12	7
2	United Technologies Corp.	1,706	145	1,777
3	Canadian Commercial Corp.	50	63	281
4	Honeywell International	453	114	713
5	MPB Corp (Timken)[a]	2	6	4
6	FAG Bearings (Precision Bearing)[a]	< 1	4	1
7	New Hampshire Ball Bearings[a]	3	8	1
8	Torrington (acquired by Timken)[a]	11	10	21
9	Jamaica Bearings	1	2	—
10	Bearing Inspection	< 1		1
11	Accurate Bushing	< 1	1	—
12	Lockheed Martin	10,223	39	6,596
13	Kaman Industrial Technologies	3	6	12
14	White Engineering Surfaces	—	1	—
15	General Electric	427	294	840

DATA SOURCE: Air Force J041 and G072D data.
NOTE: — = not applicable.
[a]Original equipment manufacturer.

CHAPTER FIVE

Summary and Conclusions

RAND conducted spend analyses for Phase I of the Purchasing and Supply Management Demonstration at the Oklahoma City ALC for purchases relating to F100 jet engines. The overall objective of the PSM demonstration was to apply commercial best practices, as recognized by research literature and practiced by innovative enterprises, in managing suppliers and the supply base to attain the best quality, performance, and prices for purchased goods and services. The demonstration was to develop a supply strategy and contract for a group of F100 requirements that would incorporate the tenets and embody the principles of PSM best practices. The Air Force selected the OC-ALC and the F100 engine for this demonstration and asked RAND to provide analytical support, including support for spend analyses.

Enterprises conduct spend analyses for three reasons. First, they can demonstrate to senior leadership how purchasing and supply management initiatives can help to achieve other goals, particularly those related to reducing overall costs or having financial resources available for other elements of the enterprise. Second, a spend analysis can help managers target specific commodity groups and specific items within those groups for purchasing and supply management initiatives. Third, a spend analysis, when conducted on an ongoing basis, can help managers develop new supply strategies. The F100 and jet engine spend analyses were used primarily to illustrate potential inefficiencies in purchasing and were used by the Air Force to select a group of requirements for the demonstration—i.e., a group of NSNs selected for applying PSM principles and tenets.

Spend analyses offer the Air Force the means to incorporate support from the supplier side to improve its supply processes. Among other things, they can highlight prospective opportunities for buyers to consolidate contracts either within firms or for specific commodities to gain leverage with suppliers, to tailor products or processes to specific needs, and to reduce transaction costs. Consolidating total business, which would also reduce transaction costs, with key suppliers would enable logistics organizations to devote more time to developing strategic relationships with their key suppliers and working on continuous supply chain improvements.

Commercial firms have long used spend analyses to uncover opportunities for improved logistics and supply processes. They often have to rely on disparate sources of data, and it is not uncommon for firms to follow the four-step process of extracting, integrating and validating, cleansing, and analyzing data, such as we describe in Chapter Two. The process is iterative, and often leads to identification of issues to be addressed in future research. This process is similar to the analysis of F100 engine spend that led to the identification of jet engine bearings as an issue worth further research. The analyses we present in this report permit two types of conclusions, the first in regard to general spend analyses for the Air Force and second in regard to the particular goods and services we examined for this study.

Regarding spend analyses in general for the Air Force, much of our effort was shaped by the fact that at the time of the F100 demonstration there was no single database that could provide information on what the Air Force purchases, who within the Air Force is buying goods and services, who the suppliers are, and what future needs will be. The F100 demonstration highlighted potential data sources to use for future spend analyses and helped to build the case for constructing an analytical tool for such analyses. We described some of the issues involved in building a composite, first-of-a-kind database of all spending data, including detailed purchase data, for the demonstration. Since the demonstration, the Air Force built the Strategic Sourcing Analysis Tool to facilitate future analyses. Such tools will need to evolve as the type and quality of data evolve.

Although the databases we used for the spend analysis were the best available for this purpose, they were not designed specifically for this use. Although they are not ideal, they offer several insights into the Air Force's spending on particular goods and services, including purchases for F100 engines, which represent a considerable expense for the Air Force, and spending for jet engine bearings, a critical commodity in maintenance of engines such as the F100. The F100 demonstration provided the Air Force with information for the first time on all purchases made for the F100 engine, including DLA purchases for F100 parts, key suppliers and their relative rank as suppliers, numbers of contracts, and types of goods and services purchased.

Both the Air Force and DLA purchased goods and services for the F100 engine. Over the period studied (FY 1999–2002), the Air Force spent more than DLA on these goods and services through a comparatively small number of contracts. Most of the Air Force dollars and contracts for the F100 engine were for items to sustain the F100 engine (rather than for acquisition of new engines), with most of those purchases being made by ALCs. Because most ALC F100 repair dollars were for the Pratt & Whitney TSS contract supporting the F100-PW-229 engine, and much of the remaining ALC F100 repair dollars were for contracts to help bridge a workload transition from the San Antonio ALC to the Oklahoma City ALC, the prospective targets for PSCM improvements in repair were relatively limited.

DLA, while spending less on the F100 than did the Air Force, purchased a wider variety of items. Because it is responsible for managing consumable parts that are less complex to manufacture and cheaper to produce, DLA made greater use of competitive sources for the F100 items that it purchased.

The Air Force and DLA use many of the same suppliers for F100 items, with 12 of the top 15 suppliers of F100 items to the Air Force and DLA combined selling to both agencies. Among these 12, the Air Force has more F100 purchases, and presumably greater leverage, with eight of them. A spend analysis conducted with respect to suppliers, rather than with respect to a particular category of goods and services, would be needed to determine whether the Air Force or DLA has greater leverage with these firms. Nevertheless, the data we examined offer some indicators of specific ways that the Air Force can leverage its buying power to exact some improvements in purchasing processes, supply management, and costs. During the demonstration, this analysis of spend by company, which had never before been detailed, showed for the first time the rank order of key suppliers and their relative importance to the Air Force and DLA, and how many contracts had been written. Knowledge of the number of goods and services contracts with each supplier gives both the Air Force and DLA an initial indication of where they might want to direct their contract consolidation efforts. Mergers among the supplier companies also might afford the Air Force and DLA with opportunities to improve their relationships with suppliers.

While our overall view of F100 items indicates that the Air Force may have the greatest leverage with the largest suppliers, narrowing the spending perspective to a commodity, such as bearings for the F100 and all other jet engines, suggests that DLA may have more leverage than the Air Force with the largest suppliers for jet engine bearings. On average, over the period we studied, DLA spent more than twice as much as the Air Force for jet engine bearings. Much of the DLA spending was for spare consumable bearings, while the Air Force was responsible for more fracture- or safety-critical and, therefore, more-expensive bearings. In short, in some cases the Air Force should lead efforts to improve purchasing processes, in some cases DLA should lead the effort, and in some cases joint leadership will be needed.

Although the Air Force and DLA purchase different types of bearings, they use many of the same suppliers. Among most of these suppliers, DLA spent more for bearings, but among some of the suppliers, the Air Force had higher total expenditures for all goods and services. This finding suggests that the Air Force may wish to work with DLA in improving its relationship with some firms, particularly bearings OEMs, but DLA may wish to work with the Air Force, or with other services, to improve its relationship with firms with which the Air Force or other service has greater spending. Both the Air Force and DLA may have some opportunities to reduce their transaction costs through consolidating their contracts with their top suppliers. Both may also consider ways to reduce their total number of suppliers to drive down their total costs. Despite the Air Force's efforts to submit part of its F100 spend to competitive contracts, most F100 contract dollars were awarded to sole-source contracts and suppliers during FYs 1999–2002, indicating that any opportunities to consolidate or otherwise improve purchasing processes with such suppliers probably lies in improving the overall supply-chain processes.

Narrowing the spending perspective to bearings, although perhaps an unforeseen choice when the Air Force was seeking to find a candidate for designing a contract using PSCM best practices (initially, the Air Force had focused on a weapon system for developing such strategies), helped to demonstrate the importance of a commodity perspective (grouping on the basis of like parts that apply across different weapon systems as opposed to grouping on the basis of equipment with a weapon system perspective) in developing supply strategies. The commodity perspective contributed to the Air Force's creation of eight commodity councils that are responsible for developing supply strategies for groups of requirements within each council's purview. Supply strategies are developed from a command-wide perspective and reviewed by AFMC headquarters. AFMC has also developed a Strategic Sourcing Analysis Tool using parts-level data to facilitate spend analyses.

Despite their limitations, existing Air Force data can be very useful for identifying opportunities for improvement in PSCM in areas of major spending, such as F100 engines, and areas of relatively small but critical spending, such as jet engine bearings. As Air Force PSCM teams gain more experience with spend analyses, they will undoubtedly uncover yet more areas for helping the Air Force get the most from its purchasing and supply resources.

References

Aberdeen Group, Inc., "Spend Visibility: Maximizing Value in Strategic Sourcing, An Executive White Paper," Boston, Mass.: Aberdeen Group, Inc., August 2002.

Amouzegar, Mahyar A., and Lionel A. Galway, *Supporting Expeditionary Aerospace Forces: Engine Maintenance Systems Evaluation (EnMasse): A User's Guide*, Santa Monica, Calif.: RAND Corporation, MR-1614-AF, 2003. Online at http://www.rand.org/publications/MR/MR1614/ (as of January 3, 2006).

Cales, Gloria, "Air Force Announces Propulsion Business Area Workload," Air Force Public Affairs, News Release, February 12, 1999. Online at http://www.defenselink.mil/releases/1999/b02121999_bt062-99.html (as of September 8, 2005).

Dearden, Dawn, "DLA Receives Overall Outstanding Small Business Program Award," DLA Public Affairs Release 03-23, July 22, 2003.

Defense Base Closure and Realignment Commission, *2005 Defense Base Closure and Realignment Commission Report,* Volume 1, 2005. Online at http://www.brac.gov/docs/final/BRACReportcomplete.pdf (as of January 3, 2006).

Dixon, Lloyd, Chad Shirley, Laura H. Baldwin, John A. Ausink, and Nancy F. Campbell, *An Assessment of Air Force Data on Contract Expenditures*, Santa Monica, Calif.: RAND Corporation, MR-274-AF, 2005. Online at http://www.rand.org/publications/MG/MG274 (as of January 3, 2006).

Dues, Timothy, "Quality Engine Development and Sustainment," Oklahoma City ALC Propulsion Product Group (OC-ALC/LR), briefing presented to the 2001 Defense Manufacturing Conference, Las Vegas, Nevada, November 26–29, 2001.

Dunn, Grover L., AF/IL, "Air Force Logistics Transformation eLog21," briefing, July 25, 2003.

"Federal Procurement and Small Business Goals: Statutory Small Business Procurement Goals for Federal Agencies," United States Small Business Administration Web site, no date. Online at http://www.sba.gov/businessop/basics/procurement.html (as of September 8, 2005).

Grimes, Jeanne, "Engine Readiness at Highest Since Gulf War," AFMC News Service Release 0327, AFMC Public Affairs Web site, March 19, 2003. Online at http://www.afmc-pub.wpafb.af.mil/HQ-AFMC/PA/news/archive/2003/Mar/0327-03.htm (as of August 10, 2005).

Heusel, Darren, "Tinker Jumpstarts Air Force-wide Supply Initiative," AFMC Public Affairs Archive, Oklahoma City Air Logistics Center Public Affairs, February 2002.

Kanakamedala, Kishore B., Glenn Ramsdell, and Paul Roche, "The Power of Purchasing Software," *The McKinsey Quarterly*, No. 4, 2003.

Mansfield, Robert E., "Improving Spares Campaign Support," *Air Force Journal of Logistics*, Vol. 26, No. 3, Fall 2002, pp. 3–9.

Miles, Donna, "DoD to Begin BRAC Closures, Realignments," *American Forces Press Service,* November 9, 2005.

Minahan, Tim A., and Mark W. Vigoroso, *The Spending Analysis Benchmark Report: Dissecting a Corporate Epidemic*, Boston, Mass.: Aberdeen Group, Inc., and Penton Media, Inc., January 2003.

Moore, Nancy Y., Laura H. Baldwin, Frank A. Camm, and Cynthia R. Cook, *Implementing Best Purchasing and Supply Management Practices: Lessons from Innovative Commercial Firms*, Santa Monica, Calif.: RAND Corporation, DB-334-AF, 2002. Online at http://www.rand.org/publications/DB/DB334/ (as of January 3, 2006).

Moore, Nancy Y., Cynthia R. Cook, Charles Lindenblatt, and Clifford A. Grammich, *Using a Spend Analysis to Help Identify Prospective Air Force Purchasing and Supply Initiatives: Summary of Selected Findings*, Santa

Monica, Calif.: RAND Corporation, DB-434-AF, 2004. Online at http://www.rand.org/publications/DB/DB434/ (as of January 3, 2006).

Porter, Anne Millen, James Carbone, Susan Avery, David Hannon, "Super Spend Analysis," *Purchasing*, Vol. 133, March 18, 2004, pp. 28–42.

Powell, CMSgt (Ret) Don, SMSgt (Ret) Dan Gray, and SMSgt (Ret) Chris Pearce, *F100 Maintenance Primer: A "How to" for Running an F100 Engine Shop*, Elmendorf AFB, Alaska, circa 1990.

Pratt & Whitney, "The F100—The Engine of Choice for 22 of the World's Best Air Forces," 2005. Online at http://www.pw.utc.com/presskit/factsheets/military_2003_f100.doc (as of August 10, 2005).

Raiff, Herman, HQ AFMC/LGPW, "Depot Maintenance Partnering," briefing, March 21, 2002.

Rukin, Karen L., "Air Force Spares Campaign," *Air Force Journal of Logistics*, Vol. 25, No. 4, Winter 2001, p. 35.

Savoie, Maj Scott, AFMC/PKL, "Enterprise-wide Strategic Sourcing: PK Director's Conference Briefing," January 29, 2003.

U.S. Air Force Deputy Chief of Staff Installations and Logistics, Directorate of Innovation and Transformation (USAF/ILI), *Future Financials: More Combat Capability for the Dollar Spent*, Washington D.C.: Pentagon, September 2004.

U.S. Air Force and Oklahoma City Air Logistics Center, *F100 Purchasing and Supply Chain Management: Phase I Report*, Tinker Air Force Base, Okla., September 6, 2002.

U.S. Air Force and Oklahoma City Air Logistics Center, *F100 Purchasing and Supply Chain Management: Phase II Report*, Tinker Air Force Base, Okla., October 6, 2003.

U.S. Air Force, Oklahoma City Air Logistics Center Supply Chain Transformation Team, "Supply Chain Council Awards for Excellence in Supply Chain Operation and Management," February 2003. Online at http://www.acq.osd.mil/log/logistics_materiel_readiness/organizations/sci/assetts/award/2003_award/2002award_winner/purchasing_and_scm_initiative.pdf (as of August 10, 2005).

Verespej, Mike, "Companies Still Learning the Value of Analyzing Data," *Purchasing*, March 17, 2005, pg. 45.

Younossi, Obaid, Mark V. Arena, Richard M. Moore, Mark A. Lorell, Jo-anna Mason, and John C. Graser, *Military Jet Engine Acquisition: Technology Basics and Cost-Estimating Methodology*, Santa Monica, Calif.: RAND Corporation, MR-1596-AF, 2002. Online at http://www.rand.org/publications/MR/MR1596/ (as of January 3, 2006).